Chinese Environmental Governance

ENVIRONMENTAL POLITICS AND THEORY

The current environmental crisis cannot be solved by technological innovation alone. The premise of this series is that the environmental challenges we face today are rooted in political crises involving political values.

Growing public consciousness about the environmental crisis and its human and nonhuman impacts, exemplified by the worldwide urgency and political activity associated with the consequences of climate change, make it imperative to study and achieve a sustainable and socially just society.

The series collects, extends, and develops ideas from the burgeoning empirical and normative scholarship spanning many disciplines with a global perspective. It addresses the need for social change from the hegemonic, consumer capitalist society in order to realize environmental sustainability and social justice.

The Series Editor is Joel Jay Kassiola, Professor of Political Science at San Francisco State University.

China's Environmental Crisis: Domestic and Global Political Impacts and Responses
Edited by Joel Jay Kassiola and Sujian Guo

Ecology and Revolution: Global Crisis and the Political Challenge
By Carl E. Boggs

Democratic Ideals and the Politicization of Nature: The Roving Life of a Feral Citizen
By Nick Garside

Chinese Environmental Governance: Dynamics, Challenges, and Prospects in a Changing Society
Edited by Bingqiang Ren and Huisheng Shou

Chinese Environmental Governance
Dynamics, Challenges, and Prospects in a Changing Society

Edited by Bingqiang Ren and Huisheng Shou

CHINESE ENVIRONMENTAL GOVERNANCE
Copyright © Bingqiang Ren and Huisheng Shou, 2013.

All rights reserved.

First published in 2013 by
PALGRAVE MACMILLAN®
in the United States—a division of St. Martin's Press LLC,
175 Fifth Avenue, New York, NY 10010.

Where this book is distributed in the UK, Europe and the rest of the World, this is by Palgrave Macmillan, a division of Macmillan Publishers Limited, registered in England, company number 785998, of Houndmills, Basingstoke, Hampshire RG21 6XS.

Palgrave Macmillan is the global academic imprint of the above companies and has companies and representatives throughout the world.

Palgrave® and Macmillan® are registered trademarks in the United States, the United Kingdom, Europe and other countries.

ISBN: 978-1-137-35068-8

Library of Congress Cataloging-in-Publication Data

Chinese environmental governance : dynamics, challenges, and prospects in a changing society / edited by Bingqiang Ren and Huisheng Shou.
   pages   cm. — (Environmental politics and theory)
Includes index.
ISBN 978-1-137-35068-8 (alk. paper)
 1. Environmental policy—China. 2. Local government and environmental policy—China. 3. Environmentalism—Political aspects—China. 4. China—Environmental conditions. 5. China—Politics and government. I. Ren, Bingqiang.
GE190.C6C45 2013
363.7'0560951–dc23   2013026300

A catalogue record of the book is available from the British Library.

Design by Integra Software Services

First edition: December 2013

10 9 8 7 6 5 4 3 2 1

To
Our Families

Contents

List of Tables and Figures ix

Series Editor's Preface xi

Acknowledgments xix

1 Introduction: Dynamics, Challenges, and Opportunities in Making a Green China 1
Bingqiang Ren and Huisheng Shou

Part I Institutions and Policies in Environmental Governance

2 A Survey of Environmental Deterrence in China's Evolving Regulatory Framework 21
Anna Brettell

3 Does Cadre Turnover Help or Hinder China's Green Rise? Evidence from Shanxi Province 83
Sarah Eaton and Genia Kostka

4 Incentive Structures and Compatible Development in a Chinese Local State 111
Jianguo Chen

Part II Firms and Environmental Regulation

5 Environmental Management, Financing, and Performance in Chinese Firms: Evidence from a Nationwide Survey 125
Xiaojun Li

6 Mind the Gap: The Role of Foreign-Invested Firms in Narrowing the Implementation Gap in China's Environmental Governance 141
Phillip Stalley

Part III Environmental Movements and Public Participation

7   Environmental Protests and Local Governance in Rural China   161
    Bingqiang Ren

8   Environmental NGOs and Participative Governance: The Case of the PM2.5 Incident   175
    Zhen Lin and Yin Guan

Part IV Social and Cultural Foundations of Environmental Governance

9   Firm Management and Environmental Organizational Violence in China   191
    Gary Green and Huisheng Shou

10 Socialization and Intergenerational Change of Environmental Consciousness in China   213
    Xiaoqing Liu

11 China's Environmental Crisis and Confucianism: Proposing a Confucian Green Theory to Save the Environment   227
    Joel J. Kassiola

Contributors   243

Index   245

Tables and Figures

Tables

2.1	Regulatory tools and enforcement mechanisms	25
2.2	Enforcement mechanisms and indirect dell	26
2.3	Annual inspections/enterprises examined/environmental violations	34
3.1	Major economic and environmental targets in the 11th FYP and 12th FYP	86
3.2	Party secretaries and mayors in Datong and Xiaoyi	92
5.1	Distribution of firms by size, pollution intensity, ownership structure and export orientation in sample cities	128
5.2	Environmental management, certification, financing and performance in Chinese firms	129
7.1	Comparison of the four cases	165
10.1	Multinomial logistic regression model of public preferences for environmental governance	218
10.2	Intergenerational difference in public preferences for environmental governance	220
11.1	A critical comparison of modernism and Confucianism	234

Figures

3.1	Project prioritization matrix: The view of a promotion-seeking official	90
4.1	Incentive structures in a compatible development model	115
5.1	Environmental departments and annual environmental reports by year of establishment and first publication	131
5.2	ISO14000 and national environmental labeling by year of certification	132
8.1	Model of Green NGOs in Environmental Participation	184
9.1	Explanatory model of organizational violence in China's environmental governance	202
10.1	Environmental consciousness, age cohort, and environmental education	222

Series Editor's Preface

The Environment, Economic Growth, and Political Legitimacy: Will China Seize the Opportunity to Be a World Environmental Leader?

Zigong asked about governing.
The Master said, "Simply make sure there is sufficient food, sufficient armaments, and that you have the confidence of the common people."
Zigong said, "If sacrificing one of these three things became unavoidable, which would you sacrifice first?"
The Master replied, "I would sacrifice the armaments."
Zigong said, "If sacrificing one of the two remaining things became unavoidable, which would you sacrifice next?"
The Master replied, "I would sacrifice the food. Death has always been with us, but a state cannot stand once it has lost the confidence of the people."

Confucius, Analects, 12:7; Slingerland, 2003:128[1]

As the series editor, it is my privilege and pleasure to introduce Bingqiang Ren and Huisheng Shou's edited volume *Chinese Environmental Governance: Dynamics, Challenges, and Prospects in a Changing Society*, the fourth volume to be published in the Palgrave Macmillan series "Environmental Politics and Theory" (see the list of previous publications). This volume's focus on China's environmental governance system complements the inaugural book in the series: *China's Environmental Crisis: Domestic and Global Impacts and Responses* (co-edited by Sujian Guo and me). Both works present a wide-ranging, multidisciplinary set of scholarly reflections on the increasingly portentous subject of China's environment, including the complex politics involved in this critical public policy issue and its system of environmental governance.

People across the world and every nation's leaders must confront the daunting challenges presented by a diverse and multitudinous array of environmental problems. These problems require deep and creative political theorizing along with effective governance principles and policy implementation guidelines. Nonetheless, China stands out as a focal point for such environmental studies (along with the United States). The reader of the this volume learns from the editors' Introduction (Section 3.3)[2] that as the "world's manufacturing powerhouse," China is now the number one consumer of energy in the world (surpassing the United States) and the number one global exporter (topping Germany). Furthermore, acknowledging that China is the greatest emitter of carbon dioxide and other greenhouse gases (again bypassing the United States), and that it is the world's most populous nation, it is clear that the country's environment holds significance for the world and, therefore, for environmental inquiry as well.

What happens in China with regard to its environment and its environmental governance—institutional structure, policymaking, and implementation—profoundly impacts not only the Chinese public and leaders but also all other nations and their citizens. Thus, the subject matter of this book is broadly relevant, applicable, and significant in today's world. Yet, surprisingly, there are few books on China's environment—and even fewer that go beyond sensational accounts of China's extreme environmental conditions—especially when compared to the surfeit of books on China's "rising" economically, politically, militarily, and diplomatically. The "Environmental Politics and Theory" series is, in part, written with the intention to fill this publishing gap. To that end, as series editor, I welcome and encourage all environmental researchers in "the burgeoning empirical and normative scholarship spanning many disciplines with a global perspective" (see Series Mission Statement (p. ii)), especially students of China's environment and its political responses, including its governance system—like the contributors to this volume—to consider this series as a forum for publishing their discussions.

I chose a passage from Confucius's *Analects* to serve as the epigraph for these prefatory remarks to highlight what I consider to be an important contribution of Ren and Shou's collection on China's environmental governance system. They emphasize the relation among China's environmental crisis, the resultant public protests, and the legitimacy of its current political order, specifically the "challenges" to China's environmental governance of "the conflict between economic development and environmental protection" and "large-scale environmental protests" (Sections 3.1 and 3.2). From its inception, Western political theory has accentuated the significance of the problem of political legitimacy (consider, for example, Aristotle, Machiavelli, Rousseau, and Marx), and as the epigraph illustrates, so did Confucius.[3]

As environmental political analysts living in the aftermath of the demise of the Soviet Union, the "Arab Spring" political rebellions, the Syrian Civil War, and myriad ongoing popular protests against political regimes worldwide (even as I write this, large-scale uprisings have sprung up in Turkey and Brazil against two governments previously considered secure and stable), we need to consider the possibility that worsening environmental conditions could lead to more regime instability and political systemic transformation. Who could doubt that this essential political subject of political legitimacy constitutes a critical issue to be examined for a formative superpower like China?

Editors Ren and Shou begin their Introduction trenchantly when they write:

> Just as the haze hangs over the city of Beijing and elsewhere in China, environmental pollution looms in the public mind across the nation. Environmental pollution not only impairs human health but also can shake the legitimacy of a government... While economic growth has become an important source of government legitimacy in China, it has also led to serious environmental pollution... The public complains about the government's economic development mode that disregards environmental protection, questions the government's capacity to make and implement sound policies, and sometimes resorts to violent protests. Public attitudes and behaviors have exerted great pressure on the government (Introduction).

These profound observations and analyses provide a perspective with which to understand the response of China's political system to the legitimacy problem created by deteriorating environmental conditions. The contributors to the volume examine and interpret how China's worsening environment forces its leaders to make changes in the nation's environmental governance in order to forestall potentially destabilizing public reactions to environmental disasters (Section 1). In short, the editors and contributors to this volume contend that in China—and significantly not in the United States at this time—the environmental crisis is a national security problem.

Considering the upgrade to full ministerial rank of the central agency charged with protecting the environment—the Ministry of Environmental Protection—and allocating it increased resources and staff, changing the law and environmental regulations as well as methods of implementation (Sections 1.1–1.3), it appears that the Chinese governmental elite—to their credit—is responding to the unsettling social reality of China's environmental crisis. Only time will tell whether these changes will be sufficient to meet the urgent task of improving China's environment and preventing popular protests. The editors assert both in the book title and in Introduction that China is a "changing society," and they want us to view this environmental challenge as an opportunity for the Chinese government. This is consistent with their "not so gloomy" (Introduction) view of the Chinese environmental future: "The [Chinese environmental] system is a complex, fluid and often self-contradictory one that defies simple labeling" (Introduction).

One is reminded of the Chinese equivalent for the English word "crisis," meaning "a dangerous opportunity for improvement." A social theorist elucidates this central concept of "crisis" as follows: "Crisis does not mean end. On the contrary, 'crisis' refers to the critical time during which the end will be avoided through new adaptations if possible, only failing these, the end becomes unavoidable."[4] The editors neither dismiss the severity of environmental crisis in China, nor do they take a gloomy view of any impending doom. Thus, the Chinese understanding of "crisis" supports their perspective regarding China's environment: China is in crisis, to be sure, but there is still time for "new adaptations" to avoid the worst end.

Creativity and courage must be called upon during crises to make the changes necessary to avoid catastrophe. I shall return to this point shortly, but now I want to focus on the central challenge to the Chinese environmental governance system described in the volume, the well-known conflict between economic growth and environmental protection.

This fundamental conflict lurks in the background of discussions in many of the chapters; for example, the desire for economic growth by Chinese local governments and cadres undermines environmental protection efforts made by Beijing. Insightfully, Ren and Shou link this central political conflict directly to the government's legitimacy.

- The difficulty to choose between the two goals [economic growth vs. environmental protection] puts the government into a plight regarding its legitimacy. Economic growth has been critical for the legitimacy of the Chinese government.
- Only by maintaining a steady economic growth rate will the government win public recognition. However, environmental protection and its impact on public health will inevitably threaten the [government's] legitimacy (Section 3.1).

I strongly agree with this analysis of the Chinese government's dilemma. Its legitimacy depends upon continuous high economic growth (China has averaged about 10 percent per year for the past 30 years and only declined to about 7.5 percent in 2012 while most of the developed Western world was struggling to reach 3 percent annual growth rate). Yet, such rapid economic growth inevitably produces environmental impairments that cause public health and media-reported disasters. These, in turn, trigger large environmental public protests undermining the government's legitimacy by exposing its incapacity to protect the health and ensure the welfare of its citizens, resulting in loss of public confidence in the state (Section 3.2).

Should the government opt to reduce economic growth in favor of protecting the environment, the resultant consequences of increased unemployment, corporate failures, loss of budgetary flexibility for social welfare or infrastructure expenditures, and all the attendant social ill-effects will likewise erode the legitimacy of the state. Therefore, political legitimacy is adversely affected no matter which horn of the legitimacy dilemma is chosen: policy option between economic growth and environmental protection produces the loss of public support for the government—a lose–lose proposition. *Might the solution to this hegemonic problem of growth versus the environment be to reject this dangerous dichotomy by giving up the value of unlimited economic growth?*

The obvious but politically unpalatable reply to this question is to recognize the undesirability of unlimited economic growth as a social goal (even though it is the heart of the neoliberal consumer capitalist order) and to explore a no-growth policy (Section 4). In the case of countries like China, with high rates of poverty, it will be necessary to pursue a policy of selective or segmented economic growth for the poor to bring them up to the level of meeting their basic needs while creating a greener and more socially just country.

My addition to this profound point is to note that China is not alone in the predicament wherein growth versus environment impacts legitimacy; not only are all developing nations caught in it, but also the developed ones. The environment is not recognized as an urgent social problem by the electorates in the developed nations; for example, in the United States environmental protection is seldom included among the top ten pressing social problems by the American public. Environmental disasters are more subtle in developed, individualistic capitalist societies (environmentally caused cancers are considered personal misfortunes and the individual's responsibility, not a social issue). Therefore, environmental threats to the legitimacy of a government that is implementing a policy of unlimited economic growth is much less harmful to political stability in developed nations. In such nations, resource depletion and environmental pollution are causing harm to the environment, but government's political legitimacy is not questioned as in China.

Let me amplify this crucial point about economic growth, the environment, and political legitimacy. A country's annual economic growth rate must be offset by environmentally caused losses that produce a lower net economic growth rate. Examples of this phenomenon are polluted water, which damages crops and ruins soil, resulting in decreased agricultural productivity, and the impact on human health caused by air pollution or the use of insecticides that are harmful to humans, which leads to decreased labor productivity as it impairs the workforce. Regarding China, I have seen estimates of as high as 3–5 percent of annual GDP growth rate as the social and environmental costs of achieving such elevated and unsustainable rates of economic growth (not to mention long-term effects of such costs on economic growth). Therefore, we must be aware of actual net economic growth levels and the real costs necessary to generate them; to ignore such costs—as is done in both developed and developing nations—is economically, ecologically, and humanely irresponsible.

Furthermore, environmental deterioration detracts from the political goal of governmental legitimacy, just like it does from real economic growth. The political costs caused to the stability and legitimacy of the government of China by environmental decay are as real—and perhaps even more harmful given the wider range of negative consequences of an unstable regime—as are the economic costs. *For a Chinese government preoccupied with social stability and political legitimacy—rightfully so given its huge population and the highest urban concentrations in the world—the environmental crisis is a clear and present political danger as well as an ecological one.* This sobering conclusion is attested to by the tens of thousands of environmental public protests in China every year (see Section 3.1), and the central government is seeking to allocate additional resources to increase the effectiveness of its environmental governance system.

This unavoidable conflict between economic growth and protecting the environment is inherent in our neoliberal consumer capitalist political economy's need for unceasing economic growth that is now globally dominant. However, because, for the most part, both China's public and leaders recognize the urgency of environmental protection, there is an opportunity to change the Chinese environmental governance system, or what Ren and Shou call "adjustments," in order to make a greener—and more secure—China (Section 4).

The editors use a special historic Chinese analogy in calling for "transforming the development model of high energy consumption" (what I have called "neoliberal consumer capitalism") as a "long march" (Section 3.1). Precisely because it suffers from a severely damaged environment with its citizens questioning the political order, China could become the much-needed laboratory and pioneer in changing the reigning development model that forms the normative base of the government's legitimacy. Economic growth-mania could be recognized for its fatal flaws and replaced with a sustainable set of new values and policies in China. Is there sufficient time for this transformational process, if it means "generations," as the editors state? (Section 4).

In conclusion, I would like to refer to this transformational process as an adaptation in response to the environmental crisis (see Frank's statement on the nature of social crises quoted earlier).

I am pleased to read the editors' endorsement of the need for fundamental change, which this book's discussions underscore: policies, institutions, implementation, and "governance". They expect the sympathetic reader of their book to conclude that

> ...we have for too long misunderstood our relationship to nature and misunderstood our position, our being, and our purpose in the universe. Yes, it is time, we believe, for an enlightenment movement, for China as well as for the world, to save the world and save ourselves. That movement, we believe, lies at the deepest and most fundamental level of China's environmental governance....
>
> (Section 4)

This is exactly what I address in the final chapter of this book wherein Ren and Shou call for an "enlightenment movement" to correct our costly mistakes with regard to humanity's relationship to nature and its position in the universe. Such transformational movements for changing the society at the "deepest and most fundamental level" of a nation's environmental governance requires more than exhortation and determination; there must be an alternative worldview to guide the required policy and institutional changes. To discover such normative guidance I sought intellectual resources within China's own cultural history that could make China a potential environmental leader of the world by providing an alternative developmental model and set of values to create a sustainable and just social order.

Confucian thought and its long tradition have served China for two-and-a-half millennia. What I aim to do in my discussion is to interpret Confucianism as a possible inspiration the Chinese and, by example, the world may use to construct a Confucian green theory and inform the enlightenment movement to save China, the world and ourselves. I hope my discussion will provoke others to pursue this line of inquiry into an alternative green society and environmental governance. It is no exaggeration to say that the future of the planet and all its inhabitants depends upon this effort.

Much more is at stake than the stability and duration of the People's Republic of China. The historical imperative concerning the realms of environmental danger and political legitimacy is befalling China, which struggles with both problems. Will China, by example, lead the world out of the ecological crisis facing both China and the world? We look to the current changing Chinese society, and to governance models presented by the contributors to this volume as they try to reconcile the irreconcilable: unlimited economic growth in a limited environment transformed into a new and effective environmental governance system with sustainable values at the base of the government's legitimacy. This volume illuminates the problems and challenges to a greener China and the world through the transformation of environmental governance; its accomplishment is one of many first steps in the long and difficult march ahead.

Joel Jay Kassiola

NOTES

1. By conventional notation regarding this work of Confucius's thoughts this passage would be indicated as follows: *Analects*, 12:7. The translation used is by Edward Slingerland, *Confucius: Analects with Selections from Traditional Commentaries* (Indianapolis: Hackett Publishing Co., 2003), p. 128.
2. Because of convenience and space limitations, I shall refer to sections of the editors' Introduction as well as the chapters referred to in these sections by them in parenthesis. Introductory remarks by the editors that are not included under any particular section will be referred to as "Introduction".
3. For a list of several passages in the *Analects* devoted to the subject of political governance, including the issue of legitimacy and stability of political regimes, see Wing-Tsit Chan, *A Source Book on Chinese Philosophy* (Princeton: Princeton University Press, 1963), p. 18.
4. See Andre Gunder Frank, "Crisis of Ideology and Ideology of Crisis," in Samir Amin, Giovanni Arrighi, Andre Gunder Frank and Immanuel Wallerstein, *Dynamics of Global Crisis* (New York: Monthly Review Press, 1982), p. 109.

Acknowledgments

Both editors of this volume were born in the early 1970s. For quite a while we felt privileged to be part of the generation that had the opportunity to witness the great transformation taking place in this country during our lifetime. Years later, however, it becomes heartbreaking to see the blue sky, sweet rains, and charming landscapes that enriched our childhood memories disappear. Who would dare to imagine, back in the 1970s, that being a Beijinger could be so miserable that suffering from terrible smog, days and nights, is considered part of a normal life? This is a hard fact that both of us, among millions of our Chinese fellows today, have to face. And we feel obligated, as scholars specializing in public policies and governance, to contribute to a budding social movement against the entrenched understanding of economic growth that focuses exclusively on numbers rather than on values and human wellbeing.

While working on our own projects on this issue, we discovered that academic gatherings on this issue were not given as much attention as they deserved. So we decided to take the lead and create an opportunity for ourselves and for our colleagues throughout the world. Thanks to the support of many individuals and organizations, an international conference was finally held at Beijing University of Aeronautics and Astronautics (Beihang University) in summer 2012. Some essays presented at the conference served as the foundation for this book project, and a few other scholars joined in shortly thereafter. There were more essays presented at the conference than we could include in this book. But our primary intention was to understand the dynamics and challenges within the country and provide an updated and in-depth analysis of the way China's environment is managed and contended among various actors and interests. In order to stay focused, we had to leave out many excellent essays on the international dimension of the issue.

This book could not have seen the light of day without the generous help and assistance of many individuals and organizations. We would like to thank the School of Public Administration at Beihang University and the government department and the School of Social Sciences and Humanities at Christopher Newport University for sponsoring the conference. We are particularly grateful to our Beihang colleagues and graduate students—too many to name—for their enthusiastic participation and assistance in the conference.

Our special thanks go to Dr. Joel Kassiola, the editor of Environmental Politics and Theory Series for Palgrave Macmillan, for inviting us to submit the manuscript and for helping us on every step throughout the process. Joe has been an incredible and indispensable supporter and advisor and we are deeply indebted to him.

We also thank Brian O'Connor and Scarlet Neath at Palgrave for their constant support, encouragement, and understanding. We particularly want to thank Scarlet for her incredible job in helping us through the tedious process and her extraordinary patience over our endless questions. We cannot imagine getting this book done without her guidance and assistance.

Bingqiang Ren would like to thank the financial support from the Fundamental Research Funds for the Central Universities and from Humanity and Social Science Youth foundation of the Ministry of Education of China for his research.

Finally, we thank our families. Both Ren and Shou have much in common in this regard. Both of us were fortunate to have our parents (in-laws for Shou) to be with us and help take care of our kids during the busy session. Their work was extraordinary and it is almost impossible for us to find appropriate words to express our gratitude. And both of us are blessed to have beautiful wives, Ying for Ren and Yingting for Shou. Their incredible understanding, care, and love have, in this case and always, been our enduring source of strength on our academic and personal journey. We dedicate this book particularly to their love. At last, we thank our kids for their love and their joy that cheered us up. Ren thanks his four-year-old son, Yining, for understanding when would be a good time to ask dad to play with him and when to leave dad alone for his work. The same goes from Shou to Rheeda, a five-year-old sweetheart and a good companion for her dad. When the book project kicked off in the late summer of 2012, Rheeda was already growing into a beautiful big girl and her little brother, Kaiden, was on the way to this world. And Kaiden came almost the same time when the book contract was signed. He grew much faster into a cheerful and energetic lad than the book project was progressing. Shou will always regret not having much time to play with him during this important period. Nevertheless, his smile and laughter, his budding teeth, and his late-night hungry cry have been a source of hope and a constant reminder that life is precious. As we state in the introduction of this book, environmental governance, in a deep sense, is not just about our wellbeing but about values and purposes, about life, about our being in this universe. We will be gone. But life goes on. And our children will live the world we hand over to them. As parents, we thank our children for teaching us how important it is to take care of this world, our families, and ourselves.

CHAPTER 1

INTRODUCTION: DYNAMICS, CHALLENGES, AND OPPORTUNITIES IN MAKING A GREEN CHINA

Bingqiang Ren and Huisheng Shou

During the rush hour commute, a haze envelops the whole city, with no trace of the sun; tall buildings nearby are obscured by the smog, and vehicles stuck in the traffic jam snail ahead. In the dusky haze, one can easily develop feelings of depression and even despair. This has been a typical picture of haze in Beijing, the capital of China. Yet it was not until the winter of 2011 that the city began to suffer from constant smoggy and hazy weather, with the air so contaminated that the pollution levels were way off the charts (more details in the chapter by Lin and Guan in this volume). Just as the haze hangs over the city of Beijing and elsewhere in China, environmental pollution looms in the public mind across the nation.

Environmental pollution not only impairs human health but also can shake the legitimacy of a government. Since instituting economic reforms and opening to the outside world, China has greatly improved its people's living standards, with rapid and sustained economic growth and upgrading of the scale and level of the economy. While economic growth has become an important source of government legitimacy in China, it has also led to serious environmental pollution. Air, water, and land are all severely polluted, posing a threat to public health. The public complains about the government's economic development mode that disregards environmental protection, questions the government's capacity to make and implement sound policies, and sometimes resorts to violent protests. Public attitudes and behaviors have exerted great pressure on the government. We predict that Chinese government will confront intense conflicts between economic development and environmental protection for a quite some time.

Yet, it is clear that China's current environmental governance system is, as the title of this book suggests, in a critical juncture and it is dangerous to jump to any firm conclusion about what is happening now and what may happen next. The governance system is a complex, fluid, and often self-contradictory one that defies simple labeling. As readers may find out from reading the chapters in this volume, the arguments and conclusions of our contributors are not always consistent from chapter to chapter. We do not see that as the weakness of this volume. On the contrary, we recognize that as the strength of it: we would like to draw our readers' attention to these contradictions and urge them to understand the complexity of the matter and explore possibilities for better solutions.

To help understand the complexity, this introduction provides readers with a comprehensive yet very brief sketch of China's current environmental governance system. We hope that the information provided here can facilitate readers' understanding of the issues in general, as well as the specific topics addressed in each chapter in this volume. Our discussion is divided into four parts, and the last section summarizes each chapter. The first part introduces the institutional, legal, and administrative framework of the governance system, as well as the regulatory regime that realizes the goals set in these frameworks. The second part discusses the problems inherent in the current system, including power distribution, decentralization, the capacity of the environmental protection agencies, and collusion between local governments and firms. The third part discusses the challenges beyond the legal and administrative mechanisms, including the contradicting goals with respect to economic growth and environmental protection inherent in China's development strategy in the past decades, as well as domestic environmental protests and the pressures introduced by globalization and the international community.

Though some of these discussions may seem depressing, we do want to make a not-so-gloomy prediction about the prospect for China's environmental future in the last part. We contend, as the book title suggests, that opportunities are available for China to improve its environment, now and in the future. As discussed in some of the chapters in this volume, various signs suggest that the country is now facing arguably the worst environmental crisis in its history; this crisis presents an opportunity to change course and create a truly sustainable development. For the government, public, and business alike, the opportunities are here and now; the legal and institutional tools from the government, the desire from the citizens, and the pressures from the world are all mature and ready.

1. Environmental Governance System: Structures and Policies

China embarked on setting up the environmental governance system in the late 1970s, when the nation was transitioning from a planned to market economy, from closure to opening to the outside world, and from centralization to decentralization of power. And this transition has profoundly affected the institutions, policies, and instruments of environmental governance in China.

1.1. Institutions for Environmental Governance

After joining the United Nations Conference on the Human Environment held in Stockholm in 1972, China began to establish its environmental governance system. In 1973, the Environment Protection Leading Group of the State Council was established as a national coordinating body of environmental policies. In 1984, the National Environmental Protection Bureau was established, becoming the environmental policy-making and executive body responsible for environmental protection and pollution prevention. A nation-wide environmental regulation system was developed, including environmental agencies at all levels as well as national environmental policies, regulations, and laws. In 1988 the bureau was granted more power after being under the direct leadership of the State Council. In 1998, the bureau was further promoted to the ministry level and became the State Environmental Protection Agency (Mol and Carter 2006:152). In 2008, it was further promoted to be the Ministry of Environmental Protection (MEP), showing that the Chinese government became more aware of the importance of environmental protection as it came under tremendous pressure due to environmental issues. MEP gained the power to participate in higher-level decision making, which enabled it to better coordinate different departments and sectors.

In addition to MEP at the center, environmental protection agencies were established at the provincial, municipal, and county levels. Local environmental protection bureaus (EPBs) implement national and local environmental policies, monitor environmental data, and impose punishments and fines for conducts violating environmental standards. Local EPBs are subject to dual leadership. They are responsible not only for the higher-level agencies (vertical level of authority) but also for the local governments at the same level (horizontal level of authority), that provide personnel and funds. Such an institutional design makes China's environmental governance operate in a grid structure, often called *tiao-kuai* (literally "branch and lump") system, which is characterized by decentralization, fragmentation, and plurality. Several chapters in this volume (Brettell; Eaton and Kostka; and Green and Shou) discuss the impact of this structure, and we will discuss it as well in detail later in this introduction.

As the environmental protection agencies are more institutionalized and their roles strengthened, their size grows too. In 1995, the staff employed by the environmental protection agencies totaled 88,000. This figure increased to 160,000 by 2004 and to 193,000 at the end of 2010 (Mol and Carter 2006). As the personnel grow in number, their capacity for policy implementation grows too, though far from a satisfactory level (Mol and Carter 2006; Stalley 2010).

1.2. Legal and Policy Framework

China has developed a legal and policy system consisting of numerous laws, regulations, and polices. In 1979, China promulgated the Environmental Protection Law (trial), the first such law. After that, China has issued a series of specialized laws on water, air, solid waste, desertification, and renewable energy. Currently, the system includes 22 laws, more than 40 regulations, around 500 standards, and more than 600 legal documents (Li 2008:118).

The establishment of the environmental governance system is characterized by "vertical development"—environmental regulatory agencies and policies are created from top to bottom by the government without much input and pressure from the public and society (Sims 1999). Since the 1970s, as China underwent drastic social changes, the government has also transformed its governance perceptions and mode, laying increasing emphasis on environmental protection and the ecological system. In 1980, environmental protection was established as the basic national policy. In 1996, the government took sustainable development as a long-term strategy. In 2003, President Hu Jintao put forward his ruling concept of "scientific development," which emphasizes comprehensive, coordinated, and sustainable development, and has since become a dominating ideology of Hu's administration. In 2012, the 18th National Congress of the Chinese Communist Party further advocated the concepts of "ecological civilization construction" and "beautiful China."

Sociopolitical transformation and changes in governmental rhetoric influence policymaking and development of environmental protection. Environmental protection development in China can be divided into three stages. The first stage, from 1972 to 1991, was primarily concerned with the control of end-of-pipe pollution. The second stage, from 1992 to 2001, focused on pollution prevention. In the third stage, from 2001 to the present, the government has taken into consideration both ecological and economic development (He et al. 2012). In sum, the focus of environmental governance has shifted from pollution control to ecological maintenance, from end-of-pipe treatment to source control, and from point treatment to regional treatment (Zhang 2008). Environmental policies and measures, as a result, have become more reasonable, playing an increasingly important role in environmental protection in China.

1.3. Environmental Regulation and Implementation

Government environmental policy implementation depends on a range of policy instruments. Under Mao, the government relied primarily on direct command–control mode to implement environmental policies. With the emergence of the market economy, economic instruments are drawing increasing attention. Economic incentives have become an important environmental governance tool in a so-called hybrid instrument that relies on multiple channels. Current environmental policy instruments can be divided into the following three types: (1) Direct supervision, which includes administrative regulation such as deadline governance and emission permits, and economic sanction such as fine for sewage discharge. (2) An incentive mechanism, which includes financial, informational, and political incentives. Financial incentives concern mainly emissions trading policies. Informational incentives are about disclosure of enterprise environmental messages and pollution sources. Political incentives, such as Rating of Model City for Environmental Protection and the Responsibility System of Environmental Protection, aim to attract attention from local government leaders to environment protection by setting environment protection as evaluation indicators of local government leaders' performance. (3) Self-regulation instruments, which mainly refers to ISO14000 certification, though it has not been fully developed yet (Li 2008:119).

From the institutional design and policymaking point of view, China has made great achievements in environmental protection. With the enhanced awareness among the government and the public, environmental protection has become one of the most burning issues in the society and has obtained higher status on the political agenda. The regulatory agencies and policy system have improved. The number of government staff for environmental protection is increasing. Environmental governance tools have become diversified.

2. Problems Associated with the Environmental Governance System

On the other hand, as Brettell's chapter in this volume discusses in detail, environmental regulatory system has not been adequate in addressing the environmental problems. In spite of numerous environmental protection agencies established at different levels and their growing size and influence, the overall trend of environmental degradation has not yet been curbed. Rather, many catastrophic environmental accidents still happen frequently. Effectiveness of environmental protection is questioned by the public. Policy implementation is often disappointing. The stark disparity between expectation and result calls for a deeper understanding of the problems in current environmental governance system. This partly can be attributed to China's rapid process of economic integration into global markets, which has drastically increased the number and size of China's manufacturing firms, the urban population, and the dependency of agriculture on pesticides and chemical fertilizers. The efforts of the government in environmental protection have thus to some extent been offset. However, the problems inherent in the environmental system itself must be taken seriously.

2.1. Power Distribution and Environmental Agencies

The power enjoyed by the environmental protection agencies is linked closely with their policymaking authority and policy-implementing capacities.

As mentioned previously and also discussed in detail by Brettell and Green and Shou in this volume, environmental protection agencies are constrained by the lack of sufficient power and authority in the current environmental protection system. First, environmental agencies, in comparison with other government branches at the same level, are lower in rank and authority, which hinders their capacity in policy implementation. For a long time, environmental agencies, as a matter of fact, had no real power and authority in relation to some other powerful departments. Before the MEP was established, environmental agencies at the center did not have the ability to fight some environment-damaging decisions made by other economic sectors. Even now, MEP still has a weaker authority compared to major economic departments such as the Development and Reform Commission, Ministry of Finance, and Ministry of Construction.

Second, legally, the governmental agencies at the same administrative level enjoy the same power and thus cannot restrain each other. This means that the MEP has no power to inhibit the administrative behaviors of each province, which is at the same administrative level; however, environmental policy implementation needs the consent and support from provinces. Similarly, it is also difficult for the ministry to restrain and control the behaviors of other ministries and state commissions, even though it often needs cooperation and support from these ministries.

Finally, local environmental agencies are subject to the leadership of the governments at the same level, and rely on the latter for fund and personnel. However, local governments pay more attention to economic growth and local officials' performance and promotion often hinge on GDP growth. Therefore, the capacity of local environmental agencies is constrained by their local peers.

In sum, policy capacity of environmental protection agencies in China to a great extent hinges on their political, rather than legal and administrative, power—greater power means higher policy implementation capability. In other words, policy implementation in China is lacking a legal foundation. As the rule of law is weak, policies are implemented not following the law, but largely driven by political power.

2.2. Consequences of Decentralization

Decentralization is an important feature of China's reform. The economic decision-making power of local governments has been enhanced over time, which has greatly advanced China's economic growth. However, decentralization has brought quite different results compared to that in the Western democracies, where decentralization enables local governments to pay more attention to the views of local citizens, so that decisions can reflect more public opinions. Under the Chinese political system, local governments have obtained great decision-making power by decentralization, but a lack of citizen monitoring has led to decision making becoming totally based on the interests and needs of local governments.

Since economic development has always been the primary goal for Chinese governments, the central government sets the targets of economic growth and expects local governments to achieve their targets. This becomes the most important element in China's cadre evaluation system and incentivizes local government officials to focus exclusively on GDP rather than environmental protection.

The fiscal system is another factor that drives local governments to ignore environmental protection. The 1994 taxation system reform strengthened the tax capacity of the central government but weakened the share for local governments, and forced them to find their own sources of revenues. Local firms—often the small and medium sized and not directly controlled by the center and therefore would pay more taxes to local governments—thus came to be important sources of revenues. Local governments actively promote local firms but ignore their environmental protection capacity. With limited capital and inadequate capacity for environmental protection, these firms often secretly discharge untreated waste water and gas in order to reduce the cost. These pollution behaviors are generally ignored and even permitted by local governments whose revenues depend on these firms.

Finally, regional differences and competitions result in lower environmental standards in China, providing possibilities for pollution transfer (for the discussion on this issue, see Chen; Stalley; and Green and Shou in this volume). Local governments are trying to promote regional economic development by inviting outside investments. In order to enhance the investment appeal of their region, local governments will keep reducing relative environmental protection standards, giving rise to the problem of pollution transfer within China. Highly polluting industries are being shifted from the economically developed coastal regions to the western areas, from urban to rural areas. Local governments of less developed regions even explicitly invite highly polluting industries to invest in their jurisdictions.

To sum, decentralization did not improve environmental quality but instead made it worse because local governments were allowed to make their own standards for environmental protection and implement (Schwartz 2004; Mol and Carter 2006). Local governments' autonomous power over development, the cadre promotion system, the taxation system, and regional competition, among other factors, make local governments inevitably motivated to choose the GDP-oriented strategy at the expense of environmental protection. With greater power and their own economic interests, local governments have strong negotiation ability against the central government, and would thus passively resist and even distort the environmental policies made by the center. Any policies contrary to the interests of local governments will be difficult to implement. Even if some of the environmental protection policies can, to a certain extent, satisfy the economic interests of local governments, their effectiveness is often uncertain because local governments can easily turn environmental policies into business activities for their own interests and benefits.

2.3. Weak Policy Implementation at the Local Level

An often discussed weakness of environmental protection in China is the weak capacity of policy implementation of the local EPBs. Their capacity depends on a variety of factors such as the power and authority of their organization, quality of their personnel, the funds available, and skills and technologies they possess. In terms of human resources, although the number of the personnel in environmental protection agencies is growing, it is on the whole still not enough, particularly at the local level. Local environmental agencies are often constrained by lack of funds. Although the central government provides funds to local environmental agencies based on projects, local governments are the primary source of funding. An additional source of funding is the penalty charges, especially the fines against polluting enterprises within their jurisdictions. This produces a perverse disincentive, hindering positive environmental protection actions (Schwartz 2004:33; Ran 2013). Local environmental agencies, taking advantage of their regulation power, often raise funds from penalty charges and substitute their own economic interests for environmental protection, thus leading to conflicts between environmental goals and economic interests. Finally, local environmental agencies severely lack pollution detection technologies and skills. While pollution in rural areas has increased rapidly, local agencies' testing skills have not improved.

2.4. Collusion between Local Governments and Industries

The interests of local governments are consistent with those of the firms. Local governments themselves play multiple roles as regulator, implementer, and a self-interest entity. Within the current institutional arrangement as mentioned, it is rational for local governments to focus on their own interests and economic growth at the expense of environmental protection. More importantly, corruption can intensify the pro-business bias of local governments. With the expanded power of administrative approval, local governments have decisive advantage over the industries under their administrative control. And yet, these powers are not monitored effectively from both above and below. Consequently, industries can easily avoid regulation of environmental policies through bribery. For both the incentive structure of the system and corruption, local governments tend to support industries. Major economic and industrial departments in local governments sometimes oppose strict environmental standards and policies (Li 2008:122) and even help industries to get away with violation of environmental regulation. Alignment of the economic interests of local government and those of firms thus hinders the effective implementation of environmental policies.

3. Challenges Facing Environmental Governance

As China is still in a phase of political, economic, and social transformation, the situation of environmental protection is extremely complex. A number of challenges are facing the environmental protection regime.

3.1. Conflicts between Economic Development and Environmental Protection

One of the most important challenges facing China is the conflict between the two goals of economic growth and environmental protection. As Chen discusses in this volume, this is a common challenge for most developing countries that are catching up economically. As technology is low and the cost for protection remains high, growth would equate environmental pollution to some extent. When most people in the western regions in China still live in poverty, environment protection does not seem to be critical in local governments' agenda. Thus unbalanced development is prevailing in many parts of China, with environmental protection being seen as secondary for both the government and the public until disasters hit the area.

The difficulty to choose between the two goals put the government into a plight regarding its legitimacy. Economic growth has been critical for the legitimacy of the Chinese government. Only by maintaining a steady economic growth rate will the government win public recognition. However, environmental pollution will inevitably threat the legitimacy. When governments' economic development goals are achieved, it tends to be difficult to satisfy the public demand for better environment. In sum, it seems inevitable that environmental protection will hinder economic development.

Yet, as discussed and demonstrated by Eaton and Kostka as well as by Chen in this volume based on the experiences of Shanxi province, this is not necessarily the case. That a compatible development model has been achieved to certain level in that region—known to be the capital of pollution in China—suggests that what matters is the institutional design and the will of the government and the society, not the stage of development. Yet it is obviously a staggering challenge for the entire country to address the conflict between the two goals and transform the development model of high energy consumption to the one with improved technology and industrial upgrading. For the vast and complex geography, it is expected to be a long march for the country to achieve this goal. The most challenging part of this effort, however, is what Ren discusses in his chapter in this volume: to realign the interests of the government and the society in order to impose real pressure on firms. That, as Ran (2013) argues forcefully, is determined to a very large extent by the central government in adjusting its goals of development and the base of its legitimacy.

3.2. Large-Scale Environmental Protests

The overall lack of institutionalized channels for the public to participate and voice their interests in environmental governance and policy implementation (Lo and Leung 2000:699–702) often leads to exacerbation of the grievance among those who suffer from fatal diseases due to polluted air and contaminated drinking water and soil. Large-scale social protests against pollution have increased rapidly. In rural areas in particular, as Ren documents in this volume, some protests often ended up with violence against the irresponsible governments and their polluting business friends. In urban areas, mass protests can be found mainly in the NIMBY (not-in-my-back-yard) movements of local residents against the construction of certain projects, such as the 2007 Xiamen protest against the construction of a chemical plant that was planned to be constructed near a residential neighborhood. More similar protests since then have taken place in various locations.

These protests have exerted great pressure upon the government, which often finds itself in a quandary: Harsh repression would damage the public's confidence in the authority and encourage firms to behave more viciously while tolerance of the protests would no doubt inspire more frequent occurrence of similar incidents in the country. Since the government is extremely concerned about social stability that social protests may undermine, environmental protests today pose a new challenge confronting the government. Multiple factors aggravate the environmental pressure and challenge the capacity of the government to handle it. Rapid urbanization results in large numbers of individuals flocking the urban area, exceeding existing capacity of cities. As individuals' living standards have been greatly improved and their lifestyle has also changed, the energy consumption per capita has been increasing drastically. The number of private cars has skyrocketed but the quality of gas remains poor because of the lack of the incentives for oil and gas companies to invest in cleaner energy. The public's environmental awareness and their desire for participation in environmental decision making are increasing but the government lacks channels and will to incorporate the public input in decisions. As Lin and Guan discuss in this volume, this is changing in recent years when the pollution crises forced the government to change their attitudes and behavior. However, given the complicated structure in the existing system as we have discussed so far, fully satisfying the public desire for participation is a daunting challenge for the government.

3.3. Globalization and the International Community

The challenges coming from the outside world are equally complicated. The process of economic integration into the world presents two types of challenges that differ in nature. The first challenge is the pressure of development, which may contribute to the worsening of environmental pollution in China. Since joining the WTO in 2001, China has been accelerating the process of globalization and has now become the world's manufacturing powerhouse. As international trade grows rapidly, the total energy consumption increases as well, placing increasing pressure on the environment. In 2010, China replaced the United States as the largest energy consumer in the world. In the same year, China became world's top exporter, passing Germany. Yet, China's exported commodities tend to be concentrated at the lower end of the chain in production and have relatively low resource utilization rate, therefore consuming more energy and resources and polluting more. Adding to that problem, the pressure of competition in the global market has been an important factor that makes the Chinese government to be reluctant in enforcing a strict environmental standard for fear of losing the edge in export. The government has long been aware of this dilemma but not until vey recent years, particularly since the eleventh five-year plan (2006–2011), did the government become firm in restructuring its industry toward a balanced one. The recent twelfth five-year plan (2011–2015) has made this transition more pronounced and concrete, giving its citizens much hope for a better environment and higher quality of life. The road to that goal, as suggested by the discussion of Brettell in this volume, is certainly a bumpy one.

On the other hand, the environmental governance regime around the world has become relatively mature and been imposing the ever greater pressure upon China. Today environmental problems have become a global issue with detrimental consequences threatening every country that is integrated into world economy. Being a country that is expected to take more international responsibilities and the one that is the most polluted, China is inevitably facing tremendous pressure from the international community to take environmental responsibility. The effect is clear. Take the global climate negotiations for example. The Chinese government has always been under the limelight in every negotiation in recent years. That pressure has forced the country to take serious steps and commit in reducing its carbon emissions by 2020 to 40–45 percent as the share of GDP of the 2005 level.

In addition to the pressures from the international community, the process of globalization itself is not entirely a "race to the bottom" as the previous paragraph indicates. Depending on circumstances, globalization may instead induce a "race to the top" if the government and the society are willing to upgrade their industries for better quality of life and possess the skills to do so. One of the critical factors for this "race to the top" is to integrate the indigenous firms into the global competition. Multinational firms are often charged for causing a "race to the bottom" and driving indigenous and local governments to be more vicious regarding environment in a crude competition. Yet, as some contributors in this volume (Stalley; Green and Shou; also Stalley 2010) and a large body of research have made clear, such a claim is unfounded. Instead, the effects of multinationals on indigenous industries' performance can be mixed (for example, Li and Stalley in this volume on Chinese firms) but the positive impact—forcing them to catch up with higher standards in order to survive in international markets—is by no means fictitious.

Again, it is a daunting task for the government as well as the firms to take this challenge, particularly when the business culture has developed into such a trenchant level of ignorance and crudity against the public wellbeing and, on the other hand, the society has gotten into the grip of this culture so hard that a cynicism has produced a public that is angry yet helpless for any meaningful collective action to come about. Green and Shou in this volume specifically address this issue and argue that this problem has aggravated to such a level that disorganization is characterizing today's Chinese society and damaging its ability to discipline its business members. Furthermore, the authors argue that two things make this issue particularly worrisome. First is the intertwined relationship and mutual enforcement between a problematic political system we have repeatedly mentioned about in this introduction and the lack of normative structure in the society that would otherwise have helped curb its members from hurting each other for financial success. Second, disorganization in a society, once in place, tends to be a generational problem.

4. Environmental Prospects: A Greener China?

Even though our tone so far has been quite critical, our answer to this question is positive and we do envision a greener China in the future. The reason is what we stated at the beginning of this introduction: the Chinese society, and its government and firms as well, are in a process of transition and the system is a complicated and fluid one with multiple possibilities. Despite all the negatives we have mentioned, many signs have emerged, suggesting a better future than one may believe. In our discussion on some problems and challenges the regime faces, we in fact have already mentioned some opportunities that are available and progresses that are made.

For example, we emphasized globalization's positive effects, which can be further confirmed by discussions in Li and Stalley in this volume. Second, we did mention that the institutions governing China's environment have been improved significantly and the environmental protection agents had never before enjoyed so much power and influence, even though much more is needed. For that matter, it really depends on the perspectives one holds when looking at the issue. From a temporal perspective, much progress has been made, even though a cross-country comparison might yield depressing findings.

At least our contributors will show some concrete progress made on the government side. Both Chen (Chapter 4) and Eaton and Kostka (Chapter 3) examine the environmental transition in Shanxi province and both chapters demonstrate that government efforts have been paid off with better designed institutions and policies. That is quite an encouraging message given that the province was once the most polluted one in the country. In particular, Easton and Kostka, by comparing two cities in the province, offer a good natural experiment that suggests that, all else being equal, institutional designs—the different way to treat the cadre turnover system—can make a difference. Indeed, even Green and Shou, though being harsh on the overall situation, acknowledge that China's gloomy future is not inevitable because it eventually depends on the willingness of the citizens to change it.

That brings us to another side of this intriguing relationship between the government and society in environmental governance. At the same time when the government is making some progresses, the societal forces are growing too. Much attention has recently been paid to the emergence of the so-called civil society out of environmental activities. The doubt on the capacity of

China's environmental non-governmental organizations (ENGOs) to change the status quo abounds. Yet, their growth is nevertheless real. A decisive factor in promoting China's ENGOs, as well as NGOs in general, is the extensive use of internet. As internet users have surged, environmental organizations have been growing rapidly in China.

Internet also improves the effectiveness of ENGOs in influencing the public and the government. Internet facilitates the capacity of ENGOs and environmentalists in educating the public for environmental awareness, mobilizing public participation in environmental protection, helping construct a green public sphere, and exerting pressure on the government (Yang and Calhoun 2007; Sima 2011). It can also make environmental information more transparent and therefore serve as channels and platforms for the public to make collective action and influence the decision making of the authority (Ren et al. forthcoming 2014). The example of the PM2.5 incident in the late 2011 and early 2012, as described by Lin and Guan in this volume, illustrates that internet and social networks associated with it can serve as an effective vehicle to facilitate the communication between the government and the public by pushing for transparency and dialogue. One may be surprised to learn from Lin and Guan that participatory governance—with public participation in every step of public policy, from policymaking to implementation to monitoring—is not just possible but is happening in China. But this case does suggest that this can happen at a right moment—when an environmental crisis threatens everyone's interests but benefits no one—with right tools: the devices monitoring PM2.5 and internet and social networking for instant communication.

The internet and social networking have another profound influence. They facilitate the process of educating the public. Both ENGOs and the government have played an important role in educating the Chinese public about environmental protection (for example, Lin and Guan on ENGOs and Liu on governmental efforts, in this volume). But the pace of this educational process has been accelerated in recent years thanks to the better communication tools. The Chinese public's rights consciousness and their interests in health and quality of life have made them more active in exerting some influence on government decision making. Whereas the traditional petition system has continued to be an important tool for the public to express their interests, other mechanisms have become increasingly important, such as public hearing, Environmental impact assessment (EIA), and the recently introduced environmental legislation that allows environmental advocate organizations to bring lawsuits against any wrongdoings. Public participation through these mechanisms no doubt will contribute to the forestalling of potential environmental risks posed by large-scale projects. In addition to empowering societal sectors, increased public environmental awareness and public participation have in fact empowered local environmental agencies as well for better policy implementation (Lo and Leung 2000).

The remaining question, however, is to what extent the Chinese public has been ready and capable of utilizing the tools available to defend their interests against a still very much reluctant government and a business group with vested interests? Green and Shou in this volume are not positive about this question, as already discussed, because restoring a normative structure in a society—a process of socialization—often takes generations. On the other hand, Liu's analysis of intergenerational change of environmental consciousness among the Chinese citizens is quite encouraging: the younger generations, particularly those born after the 1980s, show a much higher level of environmental consciousness in the sense that they, compared to their older counterparts, are more concerned about environment and more likely to see the public instead of the government as the primary party responsible for environmental protection. Liu argues that this is primarily the result of the better quality environmental education they receive before they reach adulthood. The contradicting findings in Liu and in Green and Shou might have resulted from their different focuses: whereas Liu focuses on the process of intergenerational change, Green and Shou focus more on the structure, which tends to be static and sticky. In other words, the contradiction results from one's different understanding about to what extent the so-called post-eighty generation can change the nature of Chinese environmental governance against a formidable opponent of vested interests, which the older generations have been tolerating for long.

But Liu's findings, nevertheless, are promising. They suggest that individuals can change and can be educated for good, which are agreed by Green and Shou. And our entire discussion so far also suggests that institutions, even structures, are not fixed, that they are subject to change, whether for better or for worse. The better part of our narratives—the efforts made by the government, the firms, and the public—tell us that there are good chances for China to beat the odds and change the course for a cleaner future, but fundamentally it all depends on the willingness of individuals—occupying different positions in the society but working together—to change their destiny. At the end, the most fundamental part of this effort is a better understanding of what we really want and care about. It is quite ironic that the Chinese who have suffered the most from pollution so far are those who are economically better off. A quick glance of the pollution maps widely available online makes it clear that most "cancer villages" in China today are concentrated in the eastern part of the country where GDP fares better. Perhaps, for too long, we became oblivious of what is the most important in our life or what kind of life we should value the most. Perhaps, a better understanding of what we should value the most will help changing over to a new course for a better balanced life. That is certainly not easy. The concept of "zero growth"—discussed by Chen in this volume—may upset some folks who feel so bad for the people whose living standard lags behind that of in the industrialized world. Likewise, many might consider it naive of Kossiola in this volume to call for a revival of a Confucian understanding of environment in order to address the problems inherent in the prevailing modernist understanding of Western origin. Yet, one who has so far found our discussion anything interesting and been willing to read Kossiola's chapter with an open mind should agree that we have for too long misunderstood our relationship to the nature and misunderstood our position, our being, and our purpose in this universe. Yes, it is time, we believe, for an enlightenment movement, for China as well as for the world, to save the world and save ourselves. That movement, we believe, lies at the deepest, most fundamental level of China's environmental governance, and whether China and her people are capable of grasping that opportunity will determine the prospect for a cleaner China.

5. Outline of the Book

We organize this book following our discussion in this introduction. We begin Part I with governmental rhetoric and practice in order to understand the efforts made by both the central government and the local governments, and their effectiveness. Brettell's chapter provides a detailed and updated account of China's environmental regulatory framework developed since 2006. Many specific instruments have been written into major revised or new laws, regulations, policies, and incentives and produced much potential to strengthen deterrence against lax enforcement and noncompliance. On the other hand, enforcement mechanisms have remained unsatisfactory, leaving loopholes to firms to conduct their old practices. Similarly, the mechanisms for indirect deterrence through market mechanisms and civil societies have also grown but the limitations are obvious. This disparity between the official rhetoric and practice in governmental efforts suggests, according to the author, a long way to go before China is able to create sustained and predictable deterrence and the conditions conductive to balancing economic development and environmental protection.

Although Brettell's assessment on the effectiveness of China's regulatory regime is not quite enthusiastic, other two chapters on local practices give much hope. It happens that the analysis in both chapters is based on the experiences of Shanxi province, once the most polluted region in China, in the past a few years. The difference is that Eaton and Kostka compare two cities in the province while Chen gives a broad discussion of what has happened to the entire region. Another difference is that Eaton and Kostka examine the institutional design to infer that the central government is most responsible, while Chen examines the policy designs made by the provincial government. Both chapters provide evidence that better incentives can encourage local officials to care more about environment without necessarily hurting economy. Yet the different performances in the two cities in Eaton and Kostka give a more subtle picture about the strategy and behavior of local officials at the micro level under the constraints of different institutional designs. An important finding in this study is that more locally rooted cadres can co-opt businesses into sharing the burdens of environmental policy implementation.

Part II focuses on firms' behavior. Li's analysis of the data of a 2006 nationwide survey offers an overview of the environmental performance of China's industrial enterprises—divided by location, size, pollution intensity, ownership, and export/import—from management and certification to financing and emission control. Despite great variations, the findings suggest that larger and export-oriented firms tend to comply better with environmental regulations and that state-owned large firms performed poorly. These findings echo what we have argued in this introduction, that openness, which brings in more pressure of competition and a reduced role of political connection for the success of firms, is critical to the change in the behavior of firms. That is clearly resonated in Stalley's analysis on the effect of multinational firms on the indigenous Chinese firms and on China's environmental regulation. Stalley argues that foreign investors are neither wanton polluters nor environmental stewards. Foreign investors can help narrow the implementation gap in China's environmental regulation, but their impact is largely conditional and depends on factors such as the investing firm's country of origin, industrial sector, as well as features of the host jurisdiction such as its thirst for capital and its relationship to the local communities. In other words, the host country fundamentally determines the impact of foreign firms.

Part III turns to the societal forces in environmental participation. Unlike the conventional research that tends to focus on the internal dynamics of a protest, Ren's analysis of rural environmental protests focuses on the critical role of local governance in determining the possibility of an environmental movement transforming from a peaceful and lawful negotiation to a violent confrontation with the authority. His discussion suggests that three factors in local governance determine this change: whether the local government respects the interests of residents; quality of the management of local affairs before and during the protest; the trust the local government enjoys among the local residents. Li and Guan examine an environmental movement among primarily urban residents: the PM2.5 incident in the winter of 2011 and spring of 2012. Through this movement the Chinese public forced the government to change the way it looked at environment and environmental organizations. The authors argue that the interaction between the government and environmental NGOs, although by no means smooth, eventually improved their relationship and created a model of participatory governance.

Part IV examines the social and cultural foundations underpinning public participation in environmental protection. Green and Shou offer a theoretical model to explain the incongruously high level of environmental violation by the firms in China. They argue that this results from two conditions that synergistically interact. One is the complex legal structural barriers to effective enforcement against firms' wrongdoing. The other is the socially disorganized business environment that encourages recourse to alternative codes of conduct based on the corruptive influences of a new capitalist economy supplanting the norms of traditional institutions that otherwise would exert moral pressures on individuals regarding environment.

Liu's data analysis of the two nationwide surveys, one in 1998 and the other in 2008, suggests that there has been an intergenerational difference among individuals regarding environmental consciousness. Younger generations tend to be more aware of environmental hazards and the need for environmental protection and they are more likely to see the public, rather than governments, to contribute the most in protecting environment. Liu argues that this difference in perception resulted primarily from a better quality environmental education that began from the early 1980s. The most important channel for this education is the primary and secondary schools that play a critical role for individuals in their adolescent years to be socialized into a pro-environmental culture.

Lastly, Kossiola's concluding chapter serves as a good reflection over the philosophical foundation of the rationale, strategy, institutions, and policies of China's development model in the past decades. Kossiola argues that the Confucian philosophy on the relationship between human beings and nature helps construct an intellectual framework for changing the current unsustainable model of development that dominates the world as well as China. A careful comparison between the Confucian philosophy and Western modernist philosophy regarding environment will help one to better understand the way nature should be treated and therefore the relationship between economic growth and environment.

Bibliography

He, Guizhen, Yonglong Lu, Arthur Mol, and Theo Beckers. 2012. "Changes and Challenges: China's Environmental Management in Transition." *Environmental Development* 3(complete): 25–38.

Li, Wanxin. 2008. "China's Environmental Regulation and Governance: Ideas, Capacity, Promise and Empowerment." *Public Administration Review [gonggong xingzheng yanjiu]* 5: 102–151.

Lo, Carlos Wing Hung and Sai Wing Leung. 2000. "Environmental Agency and Public Opinion in Guangzhou: The Limits of a Popular Approach to Environmental Governance." *China Quarterly* 163: 677–704.

Mol, Arther and Neil Carter. 2006. "China's Environmental Governance in Transition." *Environmental Politics* 15(02): 149–170.

Ran, Ran. 2013. "Perverse Incentive Structure and Policy Implementation Gap in China's Local Environmental Politics." *Journal of Environmental Policy & Planning* 15(1): 17–39.

Ren, Bingqiang, Huisheng Shou, and Lisheng Dong. (Forthcoming 2014). "The Urban Environmental Protests in China: Institutional Restrictions and the Internet." In Lisheng Dong, Hanspeter Kriesi, and Daniel Kübler. (eds.) *Urban Mobilization in Contemporary China*. Farnham, UK: Ashgate.

Schwartz, Jonathan. 2004. "Environmental NGOs in China: Roles and Limits." *Pacific Affairs* 77(1): 28–49.

Sima, Yangzi. 2011. "Grassroots Environmental Activism and the Internet: Constructing a Green Public Sphere in China." *Asian Studies Review* 35(4): 477–497.

Sims, Holly. 1999. "One-Fifth of the Sky: China's Environmental Stewardship." *World Development* 27(7): 1227–1245.

Stalley, Phillip. 2009. "Can Trade Green China? Participation in the Global Economy and the Environmental Performance of Chinese Firms." *Journal of Contemporary China* 18(61): 567–590.

Stalley, Phillip. 2010. *Foriegn Firms, Investment, and Environmental Regulation in the People's Republic of China*. Stanford, CA: Stanford University Press.
Yang, Guobin and Calhoun, Craig. 2007. Media, Civil Society, and the Rise of a Green Public Sphere in China. *China Information* 21(2): 211–236.
Zhang, Kun-min and Zong-guo Wen. 2008. "Review and Challenges of Policies of Environmental Protection and Sustainable Development in China." *Journal of Environmental Management* 88(4): 1249–1261.

PART I

INSTITUTIONS AND POLICIES IN ENVIRONMENTAL GOVERNANCE

CHAPTER 2

A SURVEY OF ENVIRONMENTAL DETERRENCE IN CHINA'S EVOLVING REGULATORY FRAMEWORK

Anna Brettell

1. INTRODUCTION

China has enjoyed strong economic growth for nearly three decades and has become the world's second largest economy (World Bank 2012:xxi). Part of the cost of that accomplishment has been significant and widespread industrial pollution and ecological degradation (OECD 2006:5; World Bank 2012:14).

To try to tackle these challenges, Chinese leaders took steps over the same period to develop an environmental protection regulatory framework (OECD 2006:5), but challenges remain. While some evidence suggests that the legal and policy deterrents provided for by this framework, along with technological and other advances, have contributed to positive results in some areas (Van Rooij 2010:14–37), China's environmental problems remain a considerable challenge. A variety of factors have hindered progress in tackling those challenges, including central and local authorities' prioritization of economic development (Jahiel 1994:33–52), low levels of public involvement (Brettell 2003:25), and ongoing lax implementation and enforcement and noncompliance.[1]

In and after 2006, central authorities seemingly made environmental protection more of a priority vis-à-vis other economic and social imperatives.[2] Some consider 2006 a turning point in the level of action taken by authorities to protect the environment (Wang, A 2013:1; You 2007:10837). After 2006, central authorities elevated the State Environmental Protection Administration to ministerial status making it the Ministry of Environmental Protection and added staff members;[3] and increased the number of supervision offices and personnel.[4] Central authorities also established regional environmental supervision centers (ADB 2012:96). created specialized environmental courts, expanded the number of pollutants with binding emissions reduction targets and set binding targets for energy conservation (Wang and Yan 2011:494; State Council, National Environmental Protection 11th Five-Year Plan:494). In addition, they increased overall spending on pollution control and environmental protection—although not necessarily as a percent of GDP between 2008 and 2012;[5] "shifted" from using only administrative measures in protecting the environment to also using legal, economic, and technical means (ADB 2012:92); amended or enacted new environmental laws, regulations, and policies; and expressed a stronger commitment to improve implementation and enforcement of those instruments.[6] In 2013, central Communist Party leaders revised the Party constitution to include a paragraph on "ecological civilization,"[7] signaling heightened attention at the top.

As the central leaders enhanced the priority of environmental protection, authorities enacted since 2006 new or revised environmental laws, regulations, and policies with such legal instruments as to create deterrence against noncompliance, lax implementation, and weak enforcement. In theory, the stronger the deterrents provided for by new policy and legal instruments, the higher the priority of environmental protection.

This chapter reviews some of the specific policy and legal instruments contained in new or revised Chinese laws, regulations, policies; and assesses their *potential* to improve deterrence against noncompliance, as well as their *potential* to deter lax implementation and enforcement of environmental protection directives.[8] Do revised or new Chinese laws, regulations, or policies strengthen specific and general deterrence against noncompliance, by providing for meaningful inspections and enforcement tools, environmental criminal prosecution, civil torts, administrative penalties, and other types of legal liabilities? Do new or revised laws, regulations, and policies also provide for or strengthen indirect deterrence, through market-based incentives, environmental information disclosure mechanisms, and citizen involvement in and oversight of policy and project decisions? Finally, do these instruments also provide for deterrence against lax implementation and incentives for meaningful enforcement?

This chapter will begin with a brief discussion of deterrence theory and will outline some of the legal instruments various countries employ to establish deterrence. It will also provide a brief, general overview of the main environmental regulatory and enforcement mechanisms developed in China over the past three decades. In addition, it will examine specific instruments authorities included in major revised or new laws, regulations, policies, and incentives since 2006 and discuss their potential to strengthen deterrence against noncompliance, and lax implementation and enforcement. Finally, it will discuss the significance of these new developments for understanding the potential future role for citizens in contributing to environmental deterrence, the development of the rule of law in the environmental sector, and for discerning the depth of the commitment to environmental protection on the part of central Chinese leaders.

2. DETERRENCE THEORY AND ENVIRONMENTAL REGULATORY MECHANISMS

At the core of deterrence theory is the idea that individuals or firms choose noncompliance with regulations after a cost–benefit analysis has indicated that there are greater benefits to noncompliance than with compliance (Becker 1968:169–217). In the environment sector, deterrence is successful if the cost of penalties times the probability of sanctions is higher than the costs of noncompliance. If these conditions are met, theoretically individuals or firms reportedly will choose to comply.[9] Irrespective of the deterrent effect of specific sanctions, there are also a number of economic, social, personal, managerial, and technological reasons why firms may choose compliance over noncompliance with environmental laws and policies (US EPA et al. 1992:2).

Some research found that the influence of general deterrence is as important as specific deterrence (Gray 2007: 63–84; Decker 2005: 641). Specific deterrence refers to the deterrent effects of specific regulations on the polluting behavior of any one particular plant. General deterrence refers to the deterrent effects of laws and regulations, as well as punishment of transgressors, insofar as they create a general respect for acceptable behaviors, and create the expectation in others that there will be punitive consequences for transgressions. This means that if one plant is punished for transgressions, nearby plants are more likely to get punishment for transgressions, and they will be more likely to comply with regulations.[10]

Research in other countries suggests that certain types of regulatory instruments have better deterrent effects on the behavior of polluters than others.[11] Specific instruments better for creating direct deterrence, include inspections, enforcement actions, administrative penalties, criminal prosecution, and civil torts (Shimshack 2007:26). "Rules of liability" is another specific instrument that has had a deterrent effect, at least in the United States. These types of rules stipulate a polluter may be held liable for pollution and be required to pay clean-up costs or compensation (Faure 2012:301–305). Taxes, often in the form of effluent or emission charges, have been found to act as a deterrent in China[12] and reportedly have been linked to lower environmental pollution in the United States (Faure 2012:301).

Indirect deterrence may also influence polluting behaviors (Shimshack 2007:7). Examples of indirect deterrence include transparency and freedom of expression, citizen and group action, and markets. A number of studies in China have found that the environmental performance of enterprises enrolled in China's "Green Watch" transparency program improved over time more than enterprises not enrolled in the program (Yin 2010:1). In China's Green Watch program, authorities rank the environmental performance of enterprises and make that information public. According to the Yale Environmental Performance Index, there is a positive correlation between environmental performance and the extent that citizens are allowed to participate in selecting their government, are free to express themselves, have freedom of association, and enjoy a free media (Esty et al. 2008:37).

Citizen and group action are also important. One study in America found that non-governmental organizations made a significant impact, across time and location, on the reduction of some air pollutants (Faure 2012:308–309. Faure points out that according to Public Choice Theory, optimum results are achieved when groups of various kinds, including environmental and industry groups, are competing to lobby lawmakers (Faure 2012:308). Market-based instruments, such as pollution trading permits, when used in conjunction with a cap on emissions, were found to be an efficient way to reduce pollution in some locations in the United States (Faure 2012:315–316).

Regulatory instruments reportedly will work better to deter polluting behavior and improve compliance if paired with effective implementation and enforcement practices. According to materials developed by the US Environmental Protection Agency with the Polish Environmental Ministry and the Dutch Ministry meant to help policymakers in any domestic setting, deterrence works if violators have a good chance of being detected, responses to violations are swift and predictable, sanctions are included in the response, and potential violators have the *expectation* that the first three factors are present.[13]

Chinese authorities now use a combination of legal and administrative deterrents, which based on previous research in other countries, theoretically may improve compliance as opposed to just using legal deterrents alone (Svatikova 2011:135; Faure 2012:319–320).With the use of administrative penalties, however, comes the possibility that administrative officials will act in collusion with polluters and undermine the probability of efficient deterrence (Svatikova 2011:141). In addition, previous research indicates that effective enforcement may need to be present in order for this combination of instruments to work (Faure 2012:314, 320–324).

2.1. Regulatory Tools, Enforcement, and Deterrence Mechanisms in China

Over the last few decades, Chinese leaders have gradually introduced regulatory tools and enforcement mechanisms to protect the environment. The government's responsibility to protect the environment clearly appears in the 1978 version of the PRC Constitution.[14] In 1979, the National People's Congress (NPC) passed the first environmental law, the Environmental Protection Law (Provisional) and the Environmental Protection Law (EPL) in 1989, which provides for specific environmental management tools and deterrence mechanisms.[15] Subsequent laws, regulations, measures, and policies introduced numerous other tools and mechanisms. In addition, citizens

Table 2.1 Regulatory tools and enforcement mechanisms

	Regulatory tools and action
Three simultaneous steps (design, construction, and operation of pollution prevention equipment; simultaneously with design, construction, and operation of enterprise)	Pollution discharge fees
Environmental standards	Environmental impact assessment system AND Strategic environmental impact assessment system
Monitoring • Self-reporting Computerized reporting to environmental protection bureaus (EPBs) (limited # of enterprises)	Market mechanisms • Pollution abatement subsidies • Green credit • Green insurance • Green securities • Green trade
Total load control and pollution permit system	Inspections

and groups have become more active in helping to generate deterrence. The lists on Tables 2.1 and 2.2 highlight some of these tools and mechanisms, as well as actions by citizens that have had or could have the potential to deter noncompliance and in some cases, lax implementation.[16]

3. Creating Direct Deterrence: Legal and Administrative Instruments

This section explores if and how Chinese authorities have sought to improve direct deterrence mechanisms. These include environmental criminal prosecution, civil torts, administrative penalties, fines and penalties, inspections, and other mechanisms unique to China described in the Tables 2.1 and 2.2. The focus is on mechanisms contained in laws, regulations, and measures that have some coercive power.[17]

Since 2006, authorities expanded the potential for deterrence by issuing new or revised regulatory measures covering areas that were previously unregulated. Following are several examples:

Through the PRC Tort Liability Law enacted in December 2009, authorities codified and clarified the basic principles of liability that had been scattered among various regulations (NPCSC, Tort Liability Law 2009). The Tort Law included a specific chapter on environmental liability (Tort Liability Law 2009, Chapter 8, Articles 65–68). Articles in that chapter codified into law the principle that enterprises are liable for harm caused by pollution and that if the enterprise denies the pollution is linked to the harm then it is the responsibility of the enterprise to prove that it should not

be held liable for harm. An earlier Supreme People's Court interpretation includes a similar requirement for burden of proof (Wang,C.: 97, 168). The new Circular Economy Promotion Law of the People's Republic of China, passed in August, 2008, contains several articles stipulating punishments for violations, including fines, the responsibility to cover certain costs, and criminal liability if warranted.[18]
The new Regulation on Environmental Impact Assessments for Plans issued in 2009 was the first regulation to expand the environmental impact assessment (EIA) requirement to development plans. While specific penalty amounts for infractions were not explicit in the Regulations, the Regulations extend liability to include the agencies conducting the EIA, the agencies that designate the EIA reviewing panel, and the experts and departmental representatives on the reviewing panel as opposed to just the official organizations making and approving the plan (Wang S. 2013:4).

Authorities also passed other plans and measures that extended legal liabilities into new areas, including emergency management and accident prevention.[19] They enacted the Radioactive Waste Safety Law (State Council, Radioactive Waste Safety Law), as well as the Measures for On-the-spot Inspections

Table 2.2 Enforcement mechanisms and indirect dell

Warnings	Administrative—Litigation/Mediation/Arbitration • Citizens can request admin. mediation or arbitration involving specific polluters and • Citizens can sue EPBs for violations
Criticism	Criminal sanctions—case referred to judicial organs
Administrative penalties	Civil torts—right to compensation (strict or no-fault liability)
Deadlines for action	Urban comprehensive environmental improvement fixed-quantity evaluation system (*chengshi huanjing zonghe zhengzhi dingliang kaohe zhidu*)
Injunctions (rectify problem, suspension of production, halt operations, or close facility)	Environmental quality leadership responsibility system (*Huanjing zhiliang lingdao zerenzhi*), results of which are considered in the overall evaluation of governments and officials.
Revocation of permits	Environmental protection contract (target) responsibility system (*huanbao mubiao zerenzhi*), based on policy and plan targets, provides clear targets for enterprises.
Confiscation of illegal gains or illegal goods	Administrative detention (through PSB)
Case referred to discipline inspection authorities if suspected of violation of Party or government discipline	Citizen petitions (*xinfang*)
	Non-institutionalized channels of participation
	Citizen protests and other "self-help" actions

of Pollution Source Automated Monitoring Facilities (MEP 2011:17-26)[20] amongst others.

3.1. Broader Definition of Criminal Behavior and Clarified Standards

Authorities strengthened the potential for deterrence by clarifying and lowering the threshold for classifying specified polluting behaviors as criminal through the 2011 revisions to the PRC Criminal Law. Authorities removed several requirements for designating certain types of polluting behaviors as environmental crimes in Article 338. Specifically, authorities deleted the requirements that to be a crime, the specified polluting behaviors had to cause a *"serious environmental incident"* that *"seriously pollutes the environment"* and must *"result in serious public or private property losses, or [that] cause[es] serious consequences of personal deaths and injuries"*[21] Instead, the article now stipulates that the specified polluting behavior constitutes a crime simply if it causes "serious environmental pollution." Provision 338 applies in cases where someone releases, dumps, or disposes of stipulated radioactive waste, pathogens, toxic or other dangerous waste in land, water, or atmosphere that are in violation of state stipulations.

Other stipulations in the criminal law remained the same. While authorities lowered the threshold for designating the specified polluting behaviors as criminal, they did not modify the severity of the possible prison sentences.[22] The revisions did not modify Article 408, which is relevant to the application of criminal provisions to "any functionary of a State organ who is responsible for environmental supervision and control." Criminal provisions only apply if such a functionary is in "gross neglect of duty, causes a serious accident, which results in heavy losses of public or private property or the injuries or deaths of persons."

In the spring of 2013, the Supreme People's Court and the Supreme People's Procuratorate issued a judicial interpretation on the application of law in environmental criminal cases that clarifies criteria for convictions and sentencing. The interpretation provides specific standards for applying sentences of below 3 years, between 3 and 5-7 years, or between 5-7 and 10 years in prison in specifically indicated circumstances.[23] Article one of the interpretation clarifies the definition of "serious environmental pollution" used as a criterion for applying Article 338 of the Criminal Law. It also stipulates that in addition to criminal punishments for the person directly in charge and other directly responsible parties involved in the criminal behavior, Article 6 also stipulates that the work unit involved will be fined.

The criminal statutes only help to create deterrence if they are applied. Data showed that between 2001 and 2006, there was no increase in the number of criminal cases at the national level (Van Rooij and Lo 2010:18). This trend apparently continued through 2009 as reported in the China Environment News, via People's Daily, March 9, 2009; very few people were held criminally liable for environmental crimes under the Criminal Law.

3.2. Specialized Environmental Courts, Public Interest Lawsuits, and Admission of Expert Opinion

Authorities potentially improved deterrence by expanding the number of specialized environmental courts. Authorities increased the number of specialized environmental court pilot projects from a handful in 2009 to approximately 100 by 2011 (Zhang and Zhang 2012:2). Until 2011, these courts have concentrated on "small" tasks given to them by the MEP according to one study (Huan 2011:139); they have yet to fulfill their potential.

In addition, central authorities opened the door for civil suits to be brought by a greater number of actors, although the legal foundation for such suits is still questionable. While the Marine Pollution Prevention and Control Law, an early national decision and some local court interpretations signaled authorities were experimenting with a public interest law system (Wang and Gao:37, 44–47), the 2012 Civil Procedure Law revisions was the first major law to attempt to establish a legal foundation for public interest lawsuits. Based on the Civil Procedure Law revisions, "agencies and relevant organizations stipulated by law" may file lawsuits for "acts that harm the public interest" (NPC, Civil Procedure Law 2012, Art. 55).

Creating a legal foundation for public interest lawsuits may be a major development in establishing direct deterrence, because it could potentially allow a greater number of entities to file lawsuits in cases of environmental harm. There reportedly is, however, a great deal of ambiguity as to what constitutes a public interest lawsuit and additional regulations are required to clarify which organizations may bring these suits.[24] Some of the local specialized environmental courts have already accepted suits brought jointly by environmental non-governmental organizations and government-affiliated organizations, or GONGOs.[25] Prior to this, only some government and procuratorate entities have successfully brought public interest cases to court.[26] In October 2011, a court in Yunnan province experimenting with public interest cases, accepted a lawsuit brought jointly by two non-governmental organizations (NGOs) not directly affiliated with government agencies and a local official organization involving illegal dumping of chromium sludge.[27] This marked the second time independent NGOs have participated in filing a public interest lawsuit.[28] The NGOs involved in the Yunnan case reportedly faced challenges in gathering evidence and preparing for trial, including being harassed by security guards from one of the suspected companies. The opening for public interest lawsuits to be brought by a wider number of actors may remain a development on paper if courts do not consistently accept cases. Court refusal to accept cases has been an ongoing problem that at least one expert believes is worsening.[29]

The Civil Procedure Law revisions also include other articles that may have additional deterrent effects, although they are not specifically targeted to environmental cases. Articles grant parties greater rights to request permission from the court to call on their own expert witnesses to provide "expert opinion," which could potentially assist those bringing environmental tort cases to court because they will not have to accept the conclusions of a court-appointed expert not of the party's choosing. In previous versions of the law, only the court could request relevant government departments to provide "expert conclusions."[30] Articles 152 and 154 stipulate that judicial judgments and orders must include an explanation for the court's decision; and Article 156 stipulates that court judgments and decisions should be made available to the public, except in cases involving national security, trade secrets, or individual privacy. Article 123 stipulates that a court must accept a case that fits the criteria outlined by the law, which may help with the reoccurring problem of courts refusing to accept cases.[31] It is unknown if environmental torts were a consideration when deciding to revise the relevant Articles.

3.3. Raising Fines and Increasing Costs of Noncompliance

Officials potentially improved deterrence by making illegal or below-standard polluting behaviors more expensive, following a trend between 2001 and 2006 of increasing average fines per case and a rise in the number of administrative fines.[32] Authorities reportedly increased fines for transgressions related to hazardous chemical safety in the Regulation on the Management of Hazardous Chemicals.[33] Authorities appeared to remove the ceilings for fines that could be imposed by environmental authorities at each administrative level that had been contained in several Articles in the 1999 version of the Environmental Protection Administrative Penalty Measure.[34] However, the 2009 version of the measure removed the stipulation that authorities may impose an additional fine of 3 percent of the fined amount per day that a party is overdue in paying a fine within a specified time limit.[35]

Following are some of the provisions in the Environmental Administrative Penalty amendments issued in December 2009 that could potentially raise the costs of noncompliance.[36]

If a party does not carry out an order from environmental authorities and continues with the behavior that previously incurred a punishment, then the environmental authorities may consider it as a new instance of illegal behavior, opening the door for new and additional punishments (Article 11). The 1999 version of the measure stipulated that authorities may only impose the penalty of a fine one time for the same illegal act (Article 6).

The 2009 version adds the punishment of administrative detention for law breakers (Article 7) that was not an option in the previous version.

Despite these new instruments, the 2009 revisions still stipulate that environmental authorities cannot directly issue orders for an enterprise to stop production to rectify a problem, to stop operations, or to close a facility without going through the local government (Article 16.2).

One of the most important developments to raise the costs of noncompliance has been the revision of Control Law and the Water Pollution

Prevention and Control Law in 2008. This revision increased penalties for a variety of infractions from the amounts set by the previous 1996 version of the law.

Refusing an inspection by relevant officials originally incurred fines of 10,000 Yuan or less.[37] Revised fines now range from 10,000 to 100,000 Yuan. Failing to install automatic monitoring equipment as required or failure to network with environmental protection authorities' monitoring equipment have seen similar increases in fines. (All subsequent amounts are in Yuan.)

Discharging toxic substances such as mercury, cadmium, arsenic, chromium, lead cyanide into bodies of water originally incurred fines of 10,000 or less.[38] The revised fine is not lower than 50,000 Yuan and not higher than 500,000.[39]

Placing a pollution discharge outlet in a drinking water source protection zone now incurs a fine between 500,000 and 1,000,000 Yuan (Article 75), which is the highest maximum penalty imposed by the law.

The law also increased the penalty for causing water pollution accidents by removing the cap on penalties, which was one million Yuan (1996 Law's implementing regulations, Article 43) (Li and Liu 2009). Revised fines are now equal to 20 percent of the direct losses of the incident and 30 percent in cases of serious or especially serious accidents without a cap (2008 Revisions, Article 83). The same article stipulated that the person responsible would also be held liable in addition to the facility.

Discharging waste water containing toxic pollutants or pathogens, or for storing or conveying waste water containing toxic pollutants or pathogens in ditches and other locations that do not have safeguards against seepage, originally incurred fines of less than 20,000 or 50,000 Yuan depending on the location (1996 Law's implementing regulations, Articles 39.6 and 39.7). Revised fines are now between 20,000 and 500,000 Yuan depending on the location (2008 Revisions, Articles 76.7 and 76.8).

Starting operation of a project without having water pollution prevention and control equipment built and accepted originally incurred a fine of less than 10,000 Yuan (1996 Law's implementing regulations, Article 40). The revised fines are between 50,000 and 500,000 Yuan (2008 Revisions, Article 71).

Failure to use properly, dismantling, or laying idol water pollutant facilities without approval incurred a fine of less than 10,000 Yuan (1996 Law's implementing regulations, Article 41). Revised fines are equal to 100 and 300 percent of the payable waste discharge fee, 2008 Revisions, Article 73).

3.4. Clearer and Expanded Targets

Chinese leaders reprioritized environmental protection goals and may have improved deterrence for noncompliance with environmental standards by setting clear and compulsory energy conservation and pollution reduction targets, albeit for a very limited number of pollutants. The 12th Five Year Plan includes several general environmental protection goals as well as a limited number of binding pollutant reduction targets.[40] Similar targets, expressed in absolute numbers, are included in the National Environmental Protection 12th Five Year Plan (State Council, 12th Five Year Plan, Part II, Section 3). Some of the most relevant compulsory targets for industrial pollution in the 12th Five Year Plan are outlined below.

- Reduce energy consumption per unit of GDP to 16 percent (was 20 percent at the end of 11th Five Year Plan,[41] achieved 19.06 percent of target so far).[42]
- Reduce carbon dioxide emissions to 17 percent per unit of GDP.
- Reduce total carbon oxygen demand and sulfur dioxide emissions to 8 percent (was 10 percent at the end of 11th Five Year Plan; achieved 12.45 and 14.29 percent, respectively, of target so far).
- Reduce total ammonium nitrogen discharges and nitrogen oxides emissions to 10 percent (these two pollutants were not included in the 11th Five Year Plan).

3.5. Efforts at Stronger Supervision and Enforcement

Chinese officials potentially improved deterrence by strengthening supervision and enforcement instruments. This follows a greater reliance on coercive and formal enforcement methods between 2001 and 2006 (Van Rooji and Lo 2010:18). To improve the probability of compliance, environmental authorities issued the Measures for the Ex-post Supervision of Environmental Administrative Enforcement in November 2010, which outline the requirements for environmental departments to conduct follow-up supervision examinations of the steps taken by a party that was subject to an administrative penalty action (MEP, Measures for the Ex-Post 2010, Art., 9–10).

The measure indicates that actions may be taken by environmental departments if a party refuses to carry out an administrative penalty already in force. Examples include refusing to pay levied penalties, complete required changes by a deadline, rectify illegal behaviors, stop production, stop business activities, or pay pollution fees.[43] The actions environmental departments may take range from applying to court for mandatory enforcement to applying to the relevant government agencies to order a party to suspend or halt production or close down altogether. Environmental departments may also impose penalties, dismantle equipment or facilities, or order construction suspended or stopped. In addition, they may file a report to the perpetrator's supervisory organ (such as commerce or industry and commerce departments, and the People's Bank, among others), send a case to the procuratorate for an investigation into applicable legal liabilities, or send a case to the judiciary to pursue criminal liability. Other actions may be possible depending upon the type of administrative penalty with which a party failed to comply.

Leaders also potentially improved deterrence by granting environmental and other officials more authority to take action if a party does not comply with environmental laws, regulations, or standards. Article 50 of the Administrative Coercion Law passed by the Standing Committee of the NPC on June 30, 2011, specifically stipulates that when a "party fails in its duty to fulfill an administrative decision to 'remove obstructions' or 'restoration of the status quo ante' by deadline and after a warning, and when the consequences already has or will..., create environmental pollution or destroy natural resources, an administrative agency may carry out (the decision) on behalf of the party or entrust a disinterested third party to carry it out."[44] According to Article 51, before the administrative agency carries out the coercive action, it must provide the party with notification that includes the "reason and justification" for the action, and three days prior, it must urge the party to comply with (the order). The party must "pay reasonable costs associated with carrying out (the decision) on behalf of the party, unless otherwise stipulated by law" (NPCSC, Administrative Coercion Law).[45]

The Measures on Environmental Supervision issued in 2012 provides the legal foundation for environmental officials to order the violator to rectify a pollution violation problem or rectify it within a certain frame, and issue administrative punishments if warranted (MEP, Measure on Environmental Supervision, Art. 22)

If the violation leads to serious environmental pollution or impacts society seriously, environmental authorities may make "specific demands" and "urge and supervise" relevant departments to handle legal violations within a specific time frame. In addition, environmental authorities may make the outcome open to the public (MEP, Measure on Environmental Supervision, Art. 23). If environmental responsibility targets are not completed or a serious or especially severe environmental incident occurs, environmental departments or supervision personnel have little recourse. The Measures only stipulate that they may "interview" the responsible governmental personnel at the next lower level, and may require that the government carry out its duties, implement improvement measures, and give suggestions for improving work (MEP, Measure on Environmental Supervision, Art. 25).

Environmental authorities have continued to utilize older tools to try to improve compliance including court action for failure to pay pollution levies. Utilizing the Administrative Litigation Law, authorities were able to improve collection of the levies, but there was no corresponding decrease in waste discharges or arbitrary use of authority according to one study (Zhang et al. 2010:1). In addition, authorities continued to use law enforcement campaigns to target particular types of environmental problems, industries, or legal violations. During the 11th Five Year Plan, authorities reportedly conducted five major campaigns (ADB 2012:96).

3.6. Challenges to Direct Deterrence: Lack of Political Will and Systemic Loopholes

Despite developments since 2006 described in the paragraphs above, the environmental regulatory structure in China exhibits deficiencies that could potentially continue to impede deterrence to noncompliance. Chinese leaders have yet to enact detailed laws or regulations providing guidance regarding the calculation of environmental damages and compensation amounts in environmental torts. In addition, Chinese laws do not provide for claims of punitive damages in pollution torts, which may hinder efforts to create general deterrence (Wang 2007). Chinese authorities did not include specific mention of compensation in cases of environmental damages in the 2010 revisions to the State Compensation Law of the People's Republic of China (NPCSC, State Compensation Law). Prior to 2006, however, Chinese authorities had established the legal framework supporting pollution compensation claims. Among other legal instruments, the General Principles of Civil Law, Article 124 and the Environmental Protection Law, Article 41 outline the legal framework for pollution compensation claims.[46]

Gaining consensus at the central level to strengthen environmental deterrence mechanisms has been difficult. This became apparent during efforts to draft the Environmental Protection Law, which authorities have not revised since 1989, despite numerous calls to do so over the years. In early 2011, the MEP began to draft revisions to the 1989 Environmental Protection Law, gathered suggestions from government officials, and in September 2011 sent the proposed draft revisions to the National People's Congress Environmental Protection and Natural Resources Committee (EPNRC).[47] After review, the EPNRC sent the draft to the NPC Standing Committee two months later. The NPC SC issued the draft for public comment in September 2012 for a month and collected 9,582 comments.[48] A variety of sources indicated that the draft was widely criticized.[49] Authorities reportedly will continue to revise the draft in 2013. By the time the NPC SC issued the draft for public review, such phrasings had been removed that would have provided stronger support for public participation in environmental affairs.[50] Phrasings that could have strengthened incentives conducive to improved official accountability and enforcement in the environmental sector at the local level, including making the performance of local governments more transparent, were also removed. Authorities also removed phrasings regarding official assessment of fines on a daily basis for enterprises that do not rectify illegal polluting behavior within a specified timeframe.

The difficulties in gaining consensus also were visible during the revision or drafting of other environmental laws. During the drafting stage of the Water Pollution Prevention and Control Law 2008 revisions, some authorities wanted to have higher penalties for pollution discharges into drinking water protection zones, but the penalty ceiling was limited to 1,000,000 Yuan (Li and Liu 2009). Some also supported the inclusion of daily penalties, but that provision did not make it into the final version reportedly as a result of a compromise among ministries and the business sector (Li and Liu 2009). The suggestion to include articles allowing public interest lawsuits reportedly was raised, but the resulting provision merely stated that environmental protection departments and relevant social groups may support filing lawsuits according to law those people harmed by causing water pollution (Li and Liu 2009). In the drafting phase of the Regulations on Environmental Impact Assessments for Plans, the Legislative Affairs Office of the State Council removed certain phrasings in order to reduce interdepartmental friction. The proposed phrasings stipulated that draft EIAs for plans should be disclosed to the public on a voluntary basis and that public would have a right to apply for reviewing and photocopying draft EIA reports (Wang S. 2013:6).[51]

3.7. Other Impediments to Creating Deterrence: Lack of Legislation, Environmental Authorities' Dependence upon Local Governments, and Unreliable Access to the Courts

Chinese authorities have yet to enact a major national soil pollution law, which leaves a major area of environmental concern not regulated by a major law. Authorities, however, reportedly have formed a drafting group and the MEP enacted a circular in January 2013 with plans to bring contamination survey to some sites and to bring 80 percent of those surveyed sites to standard. In addition, they would establish a monitoring network covering only 60 percent of China's arable land (Office of the State Council, Circular Regarding Work Arrangements 2013.

Inspections are an integral part of creating deterrence, and early case study research in China showed that inspections had a positive impact on the reduction of polluting behaviors (Dasgupta et al. 2000).

The numbers of inspections have varied over time with one peak in 2007, the year prior to the Beijing Olympics, followed by a drop in 2008 and a slow increase thereafter (see Table 2.3). It is unclear if annual increases will continue. Monitoring technologies in China are improving, but authorities may decrease the number of environmental inspections according to one

Table 2.3 Annual inspections/enterprises examined/environmental violations[52]

Year	Inspectors/visits	# Enterprises examined	# Environmental violations
2003–2008[53]	>7,000,000	>3,000,000	>120,000
2006	1,670,000	720,000	28,000
2007	2,300,000	1,000,000	31,000
2008	1,600,000	700,000	15,000
2009	2,420,000	980,000	10,000
2010	2,660,000	1,060,000	10,278
2011	2,700,000	1,070,000	10,000
2012	2,550,000	1,000,000	8,779

researcher, who speculated that monitoring costs were considered too high between 2006 and 2011 (Wang 2013:81).

Environmental protection bureaus (EPBs) are still too weak and dependent on local authorities. For example, in addition to being dependent upon local governments for budgets and decisions regarding personnel, some new or revised laws continue to stipulate that environmental protection authorities must apply to the local government to issue an order for an enterprise to stop production and rectify the problem, including the Water Pollution Law (Article 75). Authorities created the regional supervision centers to strengthen the vertical supervision and improve enforcement without the impediment of being dependent upon the local government.

Finally, China lacks an independent judiciary; also, it is difficult to access courts.[54] Citizens encounter a range of problems at each stage of the process when bringing environmental tort cases: courts sometimes do not accept cases, it is difficult to assess damages, winning tort cases is not easy, and enforcing judgments can be problematic.[55]

4. Indirect Deterrence and the Evolving Role of Citizens and the Market

This section looks at whether Chinese authorities have improved legal supports for indirect deterrence mechanisms, including market-based incentives, environmental information disclosure, citizen involvement in policy and project decisions, non-governmental and other organizations, and the citizen grievance system (*xinfang* system). It also examines self-help actions (citizen protests). It highlights the specific instruments included in new or revised laws, regulations, and measures, a few of which have coercive power.

4.1. Market Mechanisms Contribute to Deterrence for Big Players

After 2006, Chinese authorities unleashed a "green whirlwind" of market incentives to help reduce total discharges of major pollutants. These mechanisms included green credit, green trade, green securities, and pollution liability insurance, as described below. Authorities instituted many of these policies also as a complement to industrial policy.[56]

Green Credit: Early in July 2007, the MEP, the People's Bank of China (PBC) and the China Banking Regulatory Commission (CBRC) jointly issued the *Circular on the Implementation of Environmental Protection Policy and Regulations to Prevent Credit and Loan Risk*. The goals of the new regulation were to steer bank financing away from polluting enterprises and toward those that were more environmentally friendly.[57]

Green Securities: In 2007, environmental protection and Securities Regulatory Commission authorities issued the Circular on Environmental Protection Verification of Highly Polluting Industries to Further Standardize the Production and Management of Companies Applying for Public

Listing or Refinancing,[58] which strengthens a 2003 policy providing for Ministry of Environmental Protection input before companies in specified industries list on the stock exchange. In 2008, authorities issued additional measures to standardize and strengthen environmental evaluation of companies wishing to list on stock exchanges or making initial public offerings. These include the Guiding Opinions Regarding Strengthening Environmental Protection Supervision and Management of Listing Companies and the Public Company Environmental Protection Examination Industry Classification Management Directory.[59]

Green Export: Chinese authorities issued Circular on Enhancing the Environmental Monitoring over Export Enterprises, which stipulated that authorities would improve monitoring and increase inspections of highly polluting exporting enterprises (items 2 and 3). It stipulated that the names of enterprises that break the law will be given to the Ministry of Commerce. These companies may not receive export licenses and may not be allowed to participate in trade fairs for a specified period of time among other punishments. The names of these enterprises may also be made public.[60]

Pollution Liability Insurance: In 2007, authorities issued a guiding opinion related to pollution liability insurance, and in 2008, authorities announced that China would begin a pilot program requiring polluting industries to purchase insurance to cover costs of pollution damages and compensation to pollution victims. Authorities planned to implement the program nationwide by 2015.[61] In a step in this direction, in January 2013, authorities issued the Guiding Opinions Regarding Launching Pilot Work on Compulsory Environmental Pollution Liability Insurance, which requires enterprises in several specific industries to purchase the insurance and suggests enterprises in other industries to voluntarily purchase insurance.[62] The consequences for not purchasing the insurance include potential problems getting environmental impact assessment approvals, applying for environmental protection funds, and maintaining bank credit ratings.[63]

Other market mechanisms planned or implemented include increasing solid and water treatment charges, reforming the total load control and pollution levy systems, developing an ecological compensation system, and experimenting with tradable pollution permits (ADB 2012:107–109).

In addition, Chinese authorities began to bolster rules regarding environmental social responsibility. In 2007, the State-Owned Assets Supervision and Administration Commission issued the Guidelines to the State-owned Enterprises Directly under the Central Government on Fulfilling Corporate Social Responsibilities.[64]

These market mechanisms potentially strengthen incentives to comply with environmental laws, regulations, and policies, but only for some enterprises. Most of the mechanisms pertain only to certain industries, to exporting enterprises, to state-owned enterprises, or to large-scale enterprises.[65] In addition, establishing policies does not ensure that authorities will implement them. For example, one report cites an official study that found only 12 percent of China's 50 largest banks fully implemented their green credit policies.[66] In addition, the green credit policies of China's banks still do not influence credit ratings or lending decisions, lowering the deterrence value of the policies.

4.2. Information Disclosure Contributes to Deterrence

Since 2006, Chinese environmental and other authorities have taken steps to improve transparency in the environmental sector by creating the potential for greater citizen oversight and improved deterrence. In April 2007, the then State Environmental Protection Administration (SEPA) issued the Measures on Open Environmental Information (Measures),[67] stipulating that citizens had the right to request certain types of information about environmental affairs as outlined in posted guides and catalogues.[68] These Measures were based on the Regulations on Open Government Information (OGI Regulations) issued by the State Council in April 2007 (State Council, OGI regulations).

While the Measures were a step forward in ensuring citizens had access to information about the environment and about the activities of environmental authorities, there is no guarantee officials will release information. The Measures give officials too much discretion to refuse information requests.[69] There are many circumstances under which information would not be released;[70] for example, if requests fail to meet a recognized purpose or if the information touched on "state secrets, commercial secrets, or individual privacy."[71] In addition, there is great variation in how authorities have implemented the measures over time and location.[72] In addition, other provisions classify certain types of environmental quality data as secret, including information as outlined in the Provisions on the Scope of State Secrets in Environmental Protection Work.[73]

Authorities issued additional measures that have the potential to improve information disclosure and therefore improve deterrence.

In October 2012, the Ministry of Environmental Protection issued the Circular on Taking Steps to Strengthen Environmental Open Government Information Work.[74] If implemented, the measure may allow citizens greater access to information, especially data related to major projects of public interest, abridged versions of environmental impact assessment reports, and certain types of pollution monitoring information from "key pollution sources." The circular, however, has its shortcomings.[75]

The 2008 revisions to the Water Pollution Prevention and Control Law reportedly clarified and strengthened the legal foundation for the disclosure of standardized water quality information (Li and Liu 2009).

The 2010 Measures for the Ex-post Supervision of Environmental Administrative Enforcement stipulated that environmental departments may function within the scope of their responsibility and provide a list to the public of enterprises that refuse to carry out an administrative penalty already in force.[76]

The Environmental Administrative Penalty amendments issued in December 2009 added a new Article, which stipulated that administrative punishment decisions should be made public, with the exception of those involving state secrets, technical secrets, commercial secrets, or personal privacy.[77]

4.3. Public Participation and Deterrence

Chinese authorities have encouraged public participation in environmental policy and project decisions, while simultaneously trying to control that participation (Brettell 2003, Abstract). Citizen participation has increased over time and citizens participate through a variety of channels, including submitting comments on major policies and laws and theoretically providing input during environmental impact assessments. Citizens also join environmental social organizations, such as environmental non-government organizations— ENGOs (regardless of whether registered or non-registered), file administrative and civil lawsuits, make formal environmental complaints through the "*xinfang*" system ("letters and visits"), and utilize self-help remedies (various types of protests and demonstrations) (Brettell 2003:386–387).

4.4. Citizen Input in Environmental Impact and Social Stability Risk Assessments

After 2006, authorities improved opportunities for individuals and groups to participate in EIA processes. The contributions to deterrence, however, remain doubtful. In February 2006, authorities issued the Provisional Measures on Public Participation in Environmental Impact Assessments (2006 Provisional Measures).[78] Previously, in 2003, authorities issued a regulation stipulating that EIAs require some public input. According to the 2006 Measures, the entity conducting the EIA or other organization may arrange participation in EIAs through workshops, debates, hearings, surveys, and public hearings (MEP 2006 Measures on Public Participation, Chapter 3). Authorities later issued the Regulations on Environmental Impact Assessments for Plans, which included stipulations regarding public participation; however, the Regulations did not include information disclosure requirement (Wang S. 2013:6). The 2006 Provisional Measures only provide very general guidance regarding public participation, their implementation has been mixed, and citizens are unlikely to have opportunities to participate in a consistent and meaningful manner.[79]

Authorities drafted the Technical Guidelines for Public Participation in Environmental Impact Assessments (Comment Draft) that provide specific details about the process for public participation in EIAs; however, the MEP has not yet issued them. The MEP solicited comments on the guidelines in January 2011 (MEP, Technical Guidelines for Public Participation 2012); however, the Ministry appeared to solicit comments only from a select number of governmental departments, local governments, universities and research institutes, and quasi-governmental groups and professional associations involved in environmental protection.[80] The MEP will continue to revise the draft guidelines in 2013.

An additional requirement for "social stability risk assessments" could help to deter local authorities from approving or engaging in environmentally damaging projects that could potentially lead citizens to protest. The requirement, however, may also lead to rights abuses if authorities utilize extreme methods to ensure citizens do not jeopardize projects. In August 2012, the National Development and Reform Commission (NDRC) issued the Major Fixed Asset Investment Project Social Stability Risk Assessment Provisional Measures,[81] which directed the principles of planned large-scale fixed asset projects to investigate and analyze the social stability risk of their projects, to solicit opinions about their projects from the "relevant" public, and put forward incident prevention and resolution measures. The NDRC measure also broadly defined the three main levels of social risk and stipulated that projects with high- or mid-level risk would not be approved by the NDRC. The measure also stated enterprises and other organizations could be held liable if incidents associated with their projects caused relatively large or serious economic losses and if they had not completed a risk assessment.

4.5. Environmental Non-government Organizations

China's social organizations, including environmental organizations, vary from registered government-organized non-government organizations (GONGOs) to small, independent non-government organizations (NGOs) that do not register, or register as a business. Three main national regulations regulate registration of the broad category of "social organizations."[82] It can be difficult for groups to register because they need approval from two governmental organizations, including one that "sponsors" the group, amongst other criteria and restrictions.[83]

Over the last 20 years, despite the restrictions, environmental groups have grown in number,[84] and potentially in influence as the sector has become increasingly diverse.[85] The debate regarding the role of environmental groups in influencing policy and project decisions has not been settled, but it appears that at least some have become more active in attempting to influence agendas, policies, and projects.[86]

A few groups are proactive in exercising their right to "supervision," thereby potentially strengthening deterrence. For example, in 2006 a former journalist, Ma Jun, created water, air, and solid waste pollution maps of China and founded an NGO that has been proactive in monitoring companies that are penalized by environmental authorities, often directly writing letters to the companies to encourage them to comply with environmental laws and regulations.[87] Groups like Ma Jun's however are quite rare.

After 2006, authorities indicated that they supported the "healthy" development of non-government organizations in general, but they also appeared to strengthen monitoring and some controls over the groups. In December

2009, the State Administration of Foreign Exchange issued a notice that seemingly placed greater restrictions on foreign funding to social organizations in China.[88] According to the notice, foreign donations shall "comply with the laws and regulations...of China and shall not go against social morality or damage public interests," among many other restrictions.[89] Some believe the notice put a damper on international funding,[90] which is important because many environmental and other NGOs rely on this funding from international sources.

In December 2010, the MEP passed a guiding opinion that state authorities need to expand efforts to cultivate, "guide,"[91] and "strengthen political thought construction" (*sixiang zhengzhi jianshe*) of environmental social organizations.[92] In addition, the opinion called on authorities to further strengthen relations and cooperation between the government and social organizations,[93] and stipulated that environmental social organizations that engage in cooperative projects with foreign nongovernmental entities have to report to government officials for "examination and approval" of those activities.[94]

In March 2013, top authorities announced that by the end of the year, they would revise the three main regulations guiding the registration of social organizations. As part of those reforms, they are reportedly likely to allow industry associations, technical and scientific organizations, some charities, and community development groups to register without obtaining a "sponsor." Political, legal, religious, and foreign groups must continue to obtain a "sponsor" first before registering.[95] It is not yet clear if environmental groups that engage in advocacy or oversight work will be considered "political or legal" groups. Given this context, environmental groups may be able to help strengthen deterrence against noncompliance and against lax implementation and enforcement, but it means they may also have to contend with the limits to those efforts.

4.6. Citizen Environmental Grievances as a Deterrent: The Xinfang System

The *xinfang* ("letters and visits") system[96] has long been a legal avenue outside the judicial system for citizens to present their grievances to authorities.[97] Throughout the 1990s, environmental authorities across China worked to establish mechanisms within local environmental protection bureaus (EPBs) to accept and resolve citizens' environmental grievances.[98] Grievances can range from complaints about a neighbor's smoky furnace to very complex pollution problems that could involve death or devastation of livelihoods. Some grievances relate to cases that were not accepted by courts or could not be resolved through courts.

Some evidence suggests there were several weaknesses in the system in the 1990s through the early 2000s, including environmental officials who did not pay much attention to complaints (Brettell 2008:133–134), blunting the effectiveness of the *xinfang* system as a deterrent. In July 2006, environmental authorities issued the Environmental Petition Measure,[99] which replaced the 1997 version.[100] The 2006 version contained the requirement for EPBs to establish a "*xinfang* work responsibility system" and it stipulated that *xinfang* work would be included in annual work assessments. These requirements may have added incentives for environmental authorities to take grievances more seriously, which could add to the deterrence against noncompliance. At the same time, if petitioners took their grievance to higher-level officials, it would have a deleterious effect on the chances the lower-level official would be promoted (Minzner 2009:35).

Some studies suggest that citizens' grievances (*xinfang*) have had some impact on environmental outcomes, depending upon certain conditions.[101] According to one study, firms that were the targets of more citizen environmental grievances had less bargaining power with environmental officials to lower pollution levies (Wang H. et al. 2002).

Like citizen complaints, environmental protests in China sometimes influence the outcome of specific projects and may contribute to specific and general deterrence, as the specter of protests may deter would-be polluters. Chinese citizens have been known to take action to protest plans to build or existing pollution sources and some disputes involve violence, while others do not.[102] Since the 1970s especially, environmental disputes have been increasing (Brettell 2003:125. One 2012 article reported a 30 percent increase annually of large-scale environmental disputes. Sometimes these incidents may involve tens of thousands of people.[103] Large-scale incidents are also called "mass incidents." Authorities sometimes suppress participants and crack down on these large-scale incidents, which can involve violence leading to detentions or sentences for protesters (Brettell 2007a:167–171). Nevertheless, protests can influence the outcome of project decisions, but not always.[104]

5. Deterrence against Lax Implementation and Weak Enforcement: Progress with Limits

Implementation and enforcement of environmental laws, regulations, standards, and policies are essential steps to successful deterrence, especially given China's ongoing problems in this area. This section examines the developments since 2006 that have the potential to influence deterrence against those challenges. In particular, it examines China's interlocking set of government and official evaluation systems and new or additional provisions that stipulate legal or administrative liabilities associated with illegal or other unwanted behaviors.

5.1. Deterring Lax Implementation and Enforcement through Performance Evaluations

Central authorities developed several interlocking environmental protection assessment systems and the combined scores are likely incorporated into the main general evaluation systems[105]—that incorporated scores from all of the sector-specific assessment systems—used to evaluate the performances of governments and individual officials. While there were some early experiments that included environmental protection assessment results into the main general evaluation systems, it did not become common until 2006, when the central Party Organization Department issued the Provisional Measures Regarding Local Party and Government Leadership and Leading Cadres Comprehensive Assessment Evaluation as Required by the Scientific Development Outlook Concept.[106] Over time, authorities weighed more heavily the environmental protection evaluation scores as central authorities developed the main general evaluation systems making them more complex and sophisticated.

To meet environmental protection goals and to provide incentives for local governments and individual officials to implement the measures needed to meet these goals, central authorities first developed the environmental protection contract (target) responsibility system (*huanbao mubiao zerenzhi*) codified into law through the 1989 Environmental Protection Law.[107] This system evolved out of the Enterprise Environmental Protection Evaluation System that was started on a trial basis in 1985.[108] Authorities at all levels continue to use this system. In the contract target responsibility system, central authorities contract with provinces to meet certain environmental protection targets,[109] primarily those specified in China's five-year plans, but also targets from other plans.[110] Contracts are signed at each level down until local governments contract with individual enterprises to meet specified targets.

One local district environmental target responsibility assessment document measure reveals the weight given to various environmental indicators, explains the assessment process in general terms.[111] The measure does not specifically state the entities that conduct the annual assessment. It says "the relevant district departments organized into the environmental protection work assessment small group" use the annual assessment method to investigate and assess township governments, street offices, and industrial areas using the environmental protection target responsibility document. The Measure indicates that "district leaders" review and approve the results of the assessment, which then are sent to authorities at the next highest level and are disclosed via the government website or other means. The results are supposed to be included in the Party and government Institution leadership performance assessment work in the indicated locations.

The Measure outlines 23 indicators and their corresponding weights and indicates that authorities weigh "key" enterprises reaching emissions standards the highest at 10 points. Next, eight points are associated with each of the following three indicators: *xinfang*, leadership and department assessment, and environmental risk management. Then six points are associated with environmental propaganda and education, and administrative penalties. Investment in environmental protection is weighted quite low at two points and if investment is lower than 1.5 percent of GDP, then no points are given. Full points are given for investment over 3 percent. For *xinfang*, if the total number of complaints received is lower than the previous year, then 1 point is added to the score. If the number is higher than last year by 5 percent, then 0.02 points are subtracted. Points are subtracted if authorities do not respond to complaints in a timely fashion or if there are repeat complaints. In relation to environmental penalties, full points are given if there are no "environmental illegal behaviors." If enterprise management is "improper" (*bu daowei*), then 0.2 points are subtracted for each instance and if there is (an incident) of serious pollution to the environment, then 2 points are subtracted for each instance. Specific authorities can gain 10 "extra" credit points for "innovations" or "unique characteristics" of environmental work during the year, but the Measure does not define the criteria used to grade these "extra" point categories.

A second system authorities developed to strengthen incentives for governments in designated locations to meet certain environmental protection requirements through another assessment system is called the Quantitative Assessment System of the Comprehensive Improvement of Urban Environmental Protection (*Guanyu Chengshi Huanjing Zonghe Zhengzhi Dingliang Kaohe Zhidu de Jueding*) (Urban Assessment System). In 1988, the State Council issued the first decision regarding the system, which stipulated that 113 key environmental protection cities meet a set of specified quantitative environmental targets. By 2008, authorities expanded the number of cities subject to those requirements to 629.[112] Authorities rank these cities according to their scores for meeting specified targets and disclose that information to the public,[113] and also include those results in the environmental performance evaluations of governments and individual officials, which may spur authorities on to improve their performance. To implement the system, authorities issue five-year plans with specific targets and detailed implementation rules and regulations.[114]

The Urban Assessment System plans after 2007 included some incentives to report relevant information accurately. According to the targets and implementation rules issued in 2011 for the 12th Five Year Plan period, if authorities discover departments or responsible individuals have submitted reports that falsify, hide, fail to declare, or use an incorrect method of calculating the data, then the authorities will reduce the score for that item by 60–100 percent, depending upon the administrative level of the department.[115] Some items automatically become subject to inspection if a department claims a score of 85 percent or above for that item. Some sub-items are considered "veto items," meaning, if the department does not meet the requirements for that sub-item, then the department will not receive any points for the whole item.[116] The Quantitative Assessment System of 11th Five Year Plan issued in 2006 did not include any "veto items" nor did it include score reductions for problematic reports.[117]

The third assessment system, the Environment Quality Administrative Leadership Responsibility System (Huanjing Zhiliang Xingzheng Lingdao Fuze Zhi), is stipulated in a 1996 State Council Decision (State Council 1996).[118] Authorities instituted the system in order to ensure that local governments and main leading officials would meet specified environmental protection goals by the year 2000, including reaching pollution total load control targets, environmental protection contract (target) responsibility system targets, and other more general environmental quality conditions (State Council, 1996 and as reported in China Environment News, December 28, 1996). The Decision required local governments at each administrative level to report the results of the assessments to authorities at the next highest administrative level. While the Decision stipulated that environmental quality "will be an important component in the assessment of the work of key government leaders," it did not provide details about how that would be carried out.

In 2005, central authorities took a large step that had the potential to improve deterrence for lax implementation and enforcement by incorporating the results of the environmental protection performance assessments into the more general cadre performance evaluation assessment systems. A little background on the main performance evaluation system is needed to understand the significance of this step. Over time, the Party and the Chinese government instituted a complex set of interlocking "performance management" systems (top-down performance evaluation systems), which aim to make officials and governments more accountable and effective.[119]

These systems theoretically create incentives for better implementation and enforcement of specific Party and government directives, including environmental protection goals. The main performance evaluation systems as issued in 2009 are: the Local Party and Government Leadership and Leading Cadres Comprehensive Assessment Evaluation system, the Party and Government Operating Department Leadership and Leading Cadres Comprehensive Assessment Evaluation system, and the Party and Government Leadership and Leading Cadres Annual Assessment System.[120] These leadership and leading cadre assessment systems include numerous indicators, only a few of which relate to environmental protection. These systems link the performance of governments and officials to promotions and budget allocations, among other benefits or punishments.

Some research suggests that prior to 2005 the key determinant in evaluations of provincial-level officials was rapid GDP growth (Wu 2013:21), which may have been incentive for officials to de-prioritize environmental protection targets. Existing empirical work using provincial-level data reportedly shows that overseeing rapid GDP growth to be the most important determinant of a cadre being promoted (Li and Zhou 2005; Chen, Li and Zhou 2005). A 2009 research study found that "the current appraisal system penalizes governmental officials who adopt long-term sustainable development strategies while rewarding those who have achieved short-term GDP growth at the cost of the environment."

If China's merit-based management system encompasses city-level cadres, and if their promotion odds are higher for having overseen rapid local GDP growth, their evident preference for spending city government funds on transportation infrastructure, rather than environmental improvements (Zhao 2009:419). Authorities in some pilot projects experimented with a "green GDP" evaluation system, including Guangdong (as noted in a Southern Metropolis Daily article, November 11, 2011), but the system did not receive sufficient support from authorities, reportedly because of its potential to harm political careers, so the experiments did not spread to other locations.[121]

Therefore, in December 2005, when the State Council issued the Decision Regarding Implementing the Scientific Development Concept to Strengthen Environmental Protection,[122] central leaders were in theory sending out the message that authorities should move environmental protection higher on the list of priorities. The 2005 decision stipulates that "environmental protection is an essential component of the leadership and leading cadre assessment system, and a factor in cadre appointments and punishments and rewards" (State Council, Decision Regarding Implementing the Concept of Scientific Development, 2005, item 30). Item 30 of the decision stipulates that "local governments are required to implement environmental duties and targets through an annual target management system. Assessments of progress toward those targets are assessed regularly and the results of those assessments shall be made public." Further, item 30 stipulates that officials had to implement the "one-vote veto" system in order to achieve an "outstanding" score, although it did not specify which targets had to be met in order to avoid the veto vote. The same item also stipulates that "local governments will establish accountability systems to try to deal with the problems of local protectionism and governmental interference with the implementation of environmental laws and regulations." It stipulated that if a "serious environmental accident occurred because of policy mistakes or because leading cadres and civil servants seriously interfered in normal enforcement activities, then they will be held accountable," although it did not describe the consequences.

Central leaders also signaled to lower-level officials that environmental protection was to become a higher priority by including specific language in the most recent 12th Five Year Economic and Social Development Plan issued in 2011. That plan also illustrates that authorities further developed the main government and cadre evaluation systems and making them more important components in decisions about promotion and other rewards and punishments. Five-year plans are extremely important, as central leaders use the plans to set and achieve a set of economic and development goals. The environmental protection responsibility targets are in part based on the quantitative goals in these five-year plans.[123]

A review of the last three Chinese five-year development plans indicates central leaders gave environmental protection greater consideration in the plans' contents over time and weighed more heavily environmental performance scores in the overall performance evaluations of local governments, and later evaluations of individual officials. The Outline of the 10th Five Year Economic and Social Development Plan, issued in 2001, did not indicate that meeting the binding targets included in the plan was a factor in assessing local government performance; it just indicated authorities planned to improve assessment and supervision mechanisms.[124] The plan, however, included the binding target of reducing "key pollutants" by 10 percent.[125]

The Outline of the 11th Five Year Economic and Social Development Plan, issued in March of 2006,[126] indicated that all the binding targets included in the Plan had the "effect of law" and would be included in each department's economic and social development comprehensive evaluation and performance assessments.[127] In other words, the performance score of governments and officials in relation to meeting the binding targets in the five-year plans would now be incorporated into at least one of the three main Party and government leadership and leading cadre evaluation systems. The 11th Five Year Plan stipulated that each area shall "accept responsibility for the environmental quality in their administrative areas and implement strict environmental performance assessments, the environmental enforcement responsibility system, and accountability system."[128] In addition, it mandated the "implementation of the work unit energy consumption target responsibility and assessment system."[129]

One research project asserts that including quantitative environmental protection criteria in the list of binding targets associated with the 11th Five Year Plan contributed to the success China had in meeting the environmental protection and energy conservation goals in that plan (Wang, A. 2013:55–56). It did so indirectly by influencing the scores officials would receive for implementing the overall five-year plan, and hence influence the evaluations of local governments.

Central authorities strengthened and expanded the consequences for governments and officials of meeting five-year plan binding targets in the Outline of the 12th Five-Year Economic and Social Development Plan issued in March 2011. This most recent plan indicates that the binding targets would not only be included in the performance assessments of government institutions, but would also be included in the evaluations of individual leading cadres: "assessment results are an important basis in the determination of government leadership adjustments and the selection of officials, as well as rewards and punishments."[130] In addition, the 12th Five Year Plan stipulates that in the assessment of government institutions and officials' performances, more consideration will be given to environmental protection indicators and less will be given to indicators related to the rate of economic growth.[131]

In addition, over time, central authorities established a separate and distinct Party and government leadership accountability system (*Dang zheng lingdao ganbu wenze*), which theoretically strengthens deterrence because of the threat of being held liable for illegal behaviors or malfeasance. The 12th Five Year Plan issued in 2011, includes language to encourage the development of this system, which would be used to appraise government effectiveness and efficiency. Such assessments would not only include an internal government assessment, but would also include public "assessment through discussion" (*gongzhong pingyi*), and expert evaluation.[132] This system reportedly is to promote administrative accountability and follows previous efforts to establish similar systems. For example, in 2008, the MEP issued rules that instructed ministry personnel to actively promote both the administrative accountability system and the performance management evaluation systems, and mandated that officials clarify the scope of accountability and standardize accountability procedures, in order to raise administrative implementation efforts and increase public trust.[133] By the end of 2008, some local EPBs had issued administrative accountability system implementation rules,[134] some of which included provisions outlining the scope and procedures of the accountability system and stipulated punitive measures for violating the rules.[135] In order to spur officials into strengthening this system, in 2009, the State Council and the Office of the Party Central Committee, issued the Provisional Rules Regarding Implementing Party and Government Leader Accountability.[136]

5.2. Use of Legal and Administrative Liability Instruments to Strengthen Deterrence

To improve policy implementation and enforcement, central authorities not only developed various target responsibility and performance evaluation measures, but as time went on they also developed stipulations within new and revised laws, regulations, and policies that outlined legal and administrative punishments for a variety of illegal and undesirable behaviors by officials.

In February 2006, environmental and supervision authorities issued the Provisional Rules on Punishments for Violations of Environmental Law and Discipline to strengthen environmental protection work and to promote implementation of environmental laws and regulations. The measure outlines seven main behaviors of government administrative institutions and their officials that would result in a punishment, as well as eight main behaviors of enterprise personnel that have been appointed by state administrative organs that would result in punishment.[137] Several articles of the rules outline punishments for specified behaviors. The punishments include warnings, disciplinary action, demotion—for violations when the circumstances are relatively serious, removal from a post or dismissal (from the organization)—when the circumstances are extremely serious.[138] Other punishments outlined in the law include recording a negative mark, recording a serious negative mark, probation, and dismissal from all posts.[139]

The National Plan for Environmental Emergency Response issued in 2006 contains items providing for administrative penalties for causing environmental incidents by not consciously carrying out environmental laws and policies, and by not adhering to emergency response plans or refusing to accept the duties related to preparing for environmental emergencies, among other items.[140] According to the Interim Measures for Environmental Emergency Response Plan Management issued in 2010, environmental departments that do not draw up environmental emergency response plans, do not revise them in a timely fashion, or do not keep a plan on record according to relevant regulations, will be instructed by higher-level environmental authorities to rectify the situation (Article 24). Enterprises and organizations guilty of the same transgressions and that do not correct the situation by a given deadline will be penalized according to relevant laws (Article 25). If these transgressions result in an environmental incident or exacerbate harm, the Measures provide for penalties or criminal prosecution, if warranted, for both the responsible party and the person(s) in charge (Article 26).[141] The Measure for Environmental Emergency Information Reporting issued in 2011 provided for criticism of those failing to file reports on environmental incidents or for filing late, false, or evasive reports. It also provides for criminal prosecution of the responsible party and the person in charge if the transgression results in an environmental incident depending upon the severity and circumstances surrounding an incident.[142] The 1989 Environmental Protection Law only allowed for administrative sanction by their employer or the government.[143]

To further create deterrence against dereliction of duty that results in environmental harms, in July 2006, the Supreme People's Procuratorate issued the Provisions on the Criteria for Filing Dereliction of Duty and Rights Infringement Criminal Cases, which included a provision outlining the conditions under which authorities would pursue criminal prosecution related to a major environmental accident, if it was due to dereliction of duty.[144]

The 2008 revisions of the Water Pollution Prevention and Control Law included articles strengthening the responsibility of local officials for water quality and included incentives for them to implement and enforce laws. The Law requires officials at the county level and above to be responsible for water quality in their jurisdiction, and to include water environmental protection work in economic and social plans, as well as develop water pollution and prevention policies and measures[145] For the first time, the Law also stipulated that water environmental protection shall be included in the local environmental target responsibility system and the performance assessment and evaluation system. The score for progress toward meeting the water environmental protection targets was also included in the assessment and evaluation of both the local people's government and of relevant personnel (Article 5). Article 69 stipulates that officials could be disciplined if they fail to investigate or penalize violations or do not meet their responsibilities under the law.

The Circular Economy Promotion Law also stipulates that if a department fails to investigate violations or fails to perform its supervisory and administrative responsibilities, the government at the same or next higher level will order it to correct the failure and may impose punishments.[146]

Measures for the Ex-post Supervision of Environmental Administrative Enforcement issued in November 2010 stipulate that environmental officials responsible for supervision of compliance with administrative penalties may be subject to sanctions or criminal punishment if they shirk their responsibilities, abuse authority, or break the law for private benefit.[147] In addition, Article 13 stipulates that higher-level environmental officials may supervise the administrative penalties imposed out by lower-level environmental officials and may give feedback to lower-level environmental officials if there are problems; they may enjoin lover-level officials to take punitive actions.

They may also give feedback to lower-level people's governments or engage in joint investigations with discipline inspection or supervisory organs to pursue administrative accountability.[148] This article, however, provides only weak authority to higher-level environmental officials to directly influence lower-level environmental officials' decisions to take ex-post punitive sanctions. The threat of joint investigations with discipline inspection and supervisory organs, however, may provide indirect pressure on lower-level officials to make ex-post punitive sanctions when warranted.

In 2011, the State Council issued the Opinion Regarding Strengthening Key Environmental Protection Work, which included an item calling on governments to strengthen leadership and evaluation of environmental protection work. The item suggests that if governments have not met their environmental targets, then examination and approval processes for projects other than those related to people's livelihoods, energy conservation, and ecological and environmental protection will be temporarily halted, and relevant leaders will be held responsible.[149] The National Environmental Protection 12th Five Year Plan included similar language, and also stated that the same punitive actions would apply in cases of extremely severe environmental accidents.[150]

The Measures on Environmental Supervision issued in 2012 stipulated that environmental protection departments shall establish an evaluation system for environmental supervision personnel that rewards accomplishments and punishes behavior that violates the law or discipline. Punishments include disciplinary action, temporary or permanent seizure of law enforcement certification, or referral to judicial authorities if behavior constitutes a crime.[151]

6. Conclusion

Chinese leaders strengthened the potential for direct deterrence of noncompliance through provisions included in environmental laws, regulations, and standards in several ways: authorities clarified and expanded environmental criminal liabilities; made illegal or below-standard polluting behaviors more expensive in some cases; set a greater number of clear and compulsory energy conservation and pollution reduction targets; strengthened supervision and enforcement tools; and granted environmental and other officials more authority to take action if a party does not comply with environmental laws, regulations, or standards.

Preliminary evidence shows that this potential for direct deterrence, however, has limitations. New and revised legal and policy instruments are useful, but do they ensure that the cost of compliance is cheaper than polluting? Some evidence suggests that it is still cheaper to remain noncompliant.[152] For example, one report suggests that while environmental authorities collected an increasing amount of revenue from water pollution levies and had some success in deterring polluting behaviors, the price of these levies was still lower than abatement costs as of 2010.[153]

In addition, authorities must implement and enforce regulatory instruments to improve deterrence. Authorities increased the maximum fines for certain behaviors linked to belowe-standard water and solid waste pollution emissions, but the average fines actually imposed reportedly were fixed at nearly one-fifth of the maximum allowed by law (Faure and Zhang 2012:10037). In other words, at least in this case, authorities did not utilize fully the tools that they have at their disposal.

While authorities revised the Criminal Law so that sanctions could be applied to a wider range of activities, some argue it does not go far enough.[154] They argue behaviors that are considered criminal in other countries, such as illicit pollution emissions, are not treated as criminal under China's statutes.

Authorities also enhanced the potential for direct deterrence by building a nascent legal foundation to permit "agencies and relevant organizations stipulated by law," to bring environmental public interest cases to court. It is, however, unclear which organizations may file cases due to the vagueness of the Civil Procedure Law and the lack of other regulations to clarify the issue of legal standing. The only cases accepted by a few specialized environmental courts so far have involved government-affiliated GONGOs, or NGOs that have teamed up with environmental protection bureaus or other official organizations to jointly file a few model cases.[155] The limited number of cases and the thin legal foundation for independent NGO-filed lawsuits make it unlikely that these types of cases will change the landscape of deterrence in the foreseeable future unless authorities permit a greater number of organizations to file suits.

6.1. Leaders Strengthened the Potential for Indirect Deterrence with Less Gusto

To a limited degree, Chinese leaders also strengthened the potential for indirect deterrence of noncompliance through market mechanisms, by creating conditions for greater transparency, expanding channels of public participation in regulatory and project decision-making processes, loosening restrictions on some types of non-governmental organizations (NGOs), and to some degree creating incentives for officials to be more responsive to citizen grievances. The potential for stronger indirect deterrence, however, is even more limited than the potential authorities created for direct deterrence.

Market mechanisms only target certain industries, relatively large companies, or companies that export goods. In addition, some research illustrated that these mechanisms are not always utilized.

Authorities expanded the potential for environmental transparency by issuing the 2007 Measure on Open Environmental Information and various other provisions in major laws and regulatory instruments. These instruments may not go far enough and are too flexible, however, which allows for arbitrary implementation. And in addition other provisions continue to classify some data on environmental quality as "secret." While citizens and groups are growing more proactive in submitting information requests and in some cases receiving the information they requested, in many other cases, however, authorities do not release the information. If officials only irregularly disclose materials and companies are not required to disclose to the public pollution emissions and effluent discharge data, then the potential for indirect deterrence created by transparency initiatives will be limited.

Environmental authorities expanded the potential for public participation in environmental impact assessment processes through the 2006 provisional Measures on Public Participation in Environmental Impact Assessments and additional regulations and guidelines. They took the additional step to mandate that EIA abridged reports should be posted online. Abridged versions of the reports, however, must have sufficient information for citizens to understand the possible environmental consequences of projects or plans. Participation by the general population in environmental impact assessments for projects and plans that may have a direct impact on their lives appears to be inconsistent at best.

Authorities simplified processes for some environmental quasi-governmental and non-government organizations to register and expanded the space for them to act freely, thereby contributing to indirect deterrence. At least one quasi-government organization and one more independent organization have been active in providing legal assistance and advice to pollution victims. On the other hand, central authorities have made it harder for other groups by excluding them from simplified registration processes, by hampering or making it more difficult to have access to foreign funding, and by increasing monitoring of and "guiding" group activities. A very limited number of environmental groups have taken the initiative to monitor companies' environmental performance and to "supervise" local environmental officials. Until a greater number of organizations can act freely, the overall potential for citizen environmental groups to contribute to indirect deterrence remains relatively low.

Authorities continue to support the expression of citizen grievances and have given local officials some incentives to resolve issues locally. In some cases, however, the incentives have led to human rights abuses by officials. Nevertheless, citizen environmental grievances have continued, and likely will continue, to help create deterrence to noncompliance. Authorities encourage environmental complaints for many reasons: it helps them discover illegally polluting enterprises; the complaint system is a "safety valve" to channel dissatisfaction so that citizens do not utilize more disruptive methods to express their grievances; and complaints lower the bargaining power of polluting enterprises and strengthen the authority of environmental officials to deal with them.[156]

Citizen protests over planned or existing polluting facilities are increasing and may influence specific project decisions. The threat of protests may strengthen specific as well as general deterrence in some locations. In other locations, however, authorities may work very hard to prevent protests or suppress them when they do occur, blunting their deterrent effect. Officials get docked points on their performance evaluations for "mass incidents" or "social instability," which can be a strong incentive for officials to nip protests in the bud or suppress them.

6.2. Improved Deterrents to Lax Implementation and Enforcement: Evaluation Systems and Legal and Administrative Sanctions

Authorities bolstered deterrents to lax implementation and enforcement, especially in 2006 and after by reforming and making the three main Party and government leadership and leading cadre performance evaluation systems more important components in decisions to promote officials or give rewards and punishments. In addition, central authorities linked environmental protection performance assessment systems to the three main evaluation systems. The usefulness of these measures, however, depends upon their implementation, which remains problematic, and the weightage authorities give to environmental protection criteria in the three main general evaluation systems.

Also important are the efforts of central authorities to strengthen some of the legal and administrative liability instruments to combat corruption, malfeasance, and other behaviors, including those contained in the Provisional Rules on Punishments for Violations of Environmental Laws and Discipline. These measures, however, must be implemented with consistency to contribute to specific and general deterrence.

6.3. Challenges That Weaken Deterrence and Lack of Consensus at the Top

Despite deterrence efforts since 2006, lax implementation and noncompliance continue to blunt the effectiveness of environmental laws, regulations, standards, and policies.[157] Weaknesses inherent in China's performance evaluation systems and their implementation may be part of the reason for the continued problems. For example, while authorities have made several different environmental and ecological protection or conservation items "one-vote veto" items, many other sectors also have increased the number of "one-vote veto" items. One article points out that in one province there are 62 "one-vote veto" items,[158] which could blunt the effectiveness of the ones focused on environmental protection. The same article notes a Nanjing official criticized the problem of falsifying information in evaluations. It also points out that while one city designated energy conservation advances as a "one-vote veto" item, few officials have been fired if they do not meet requirements. Another article details the reasons for lax implementation of strict evaluation measures, including problems with goal specification, goal displacement, data accuracy and falsification, and possible collusion, among others (Wang, A. 2013:61–73)

It is unlikely that tools for direct deterrence alone are going to be enough to improve compliance and implementation, and compliance. Indirect deterrence is also needed, but in China indirect deterrence mechanisms are underdeveloped. Some research suggests that a "convergence" of various factors leads to improved enforcement, including social pressures, governmental support, and economic conditions (Van Rooji and Lo 2010:14–37). The need for a convergence of factors for better enforcement has implications for China's policy of moving polluting enterprises out of the more developed coastal areas to poorer areas within the same province, to poorer areas in other provinces, or to China's interior and western regions (OECD 2007:232–236). In poorer areas, conditions may favor noncompliance and lax implementation, environmental transparency may be lower (CECC 2012:117), there may be less public participation (Li Bo, Chinadialogue, June 18, 2013), and the capacity of environmental protection authorities may be weaker.[159]

Some signs indicate that there is still a lack of consensus at the top regarding the importance that should be granted to environmental protection, and holes remain in the regulatory structure. It has been pointed out that environmental protection agencies are still understaffed.[160] Although authorities increased investment in environmental protection overall, they have not increased investment as a percentage of GDP for several years. The lack of consensus is evident in the debates regarding draft revisions to the Environmental Protection Law and other laws. In addition, some environmental laws have not been revised for many years and the lack of a major law regulating soil contamination is a glaring hole in the regulatory system. There are potentially serious consequences for food safety and health among other issues by leaving one of the major pollution-bearing mediums, soil, without an overarching national law to protect it.

It seems clear that around 2006, central authorities elevated the priority of environmental protection to some degree and strengthened deterrence mechanisms; however, there are weaknesses in these mechanisms and their enforcement. It also seems clear that central Chinese leaders have not generated the consensus needed to strengthen deterrence enough to make the cost of compliance lower than the cost on noncompliance. Continued reliance on top-down accountability and official evaluation systems and less reliance on bottom-up supervision creates less than optimum deterrence against lax implementation and enforcement. Central leaders have a long way to go before they create sustained and predictable deterrence and the conditions conducive to balancing economic development and environmental protection, and protection of the health of Chinese citizens.

NOTES

1. For more specific information on lax implementation and enforcement, see *Congressional-Executive on China Annual Reports*, 2010, 154; 2011, 147; and 2012, 114–115; Abigail Jahiel, "Policy Implementation through Organizational Learning: The Case of Water Pollution Control in China's Reforming Socialist System," PhD Dissertation (The University of Michigan, 1994), 33–52; Wang Jin and Yan Houfu, "Barriers and Solutions

to Better Environmental Enforcement in China," *Ninth International Conference on Environmental Compliance and Enforcement Proceedings*, Whistler, British Columbia, Canada, June 20–24, 2011, 494 http://inece.org/conference/9/proceedings/56_WangYan.pdf; OECD, *Environmental Compliance and Enforcement in China: An Assessment of Current Practices and Steps Forward: An Assessment of Current Practices and Ways Forward*, draft report presented at the 2nd meeting of the Asian Environment Compliance and Enforcement Network, December 4–6, 2006, 5. http://www.oecd.org/environment/outreach/37867511.pdf; Carlos Wing-Hung Lo, Gerald E. Fryxell and Wilson Wai-Ho Wang, "Effective Regulations with Little Effect? The Antecedents of the Perceptions of Environmental Officials on Enforcement Effectiveness in China," *Environmental Management*, Vol. 38, No. 3 (2006): 388, 389, via Springer Link, http://link.springer.com/article/10.1007%2Fs00267-005-0075-8?LI=true; Carlos Wing Hung Lo and Sai Wing Leung, "Environmental Agency and Public Opinion in Guangzhou: The Limits of a Popular Approach to Environmental Governance," *The China Quarterly*, No. 163 (September 2000):677–704, http://links.jstor.org/sici?sici=0305-7410%28200009%290%3A163%3C677%3AEAAPOI%3E2.0.CO%3B2-L. Anna Brettell, "The Politics of Public Participation and the Emergence of Environmental Proto-Movements in China," PhD Dissertation (University of Maryland, 2003), 25, http://drum.lib.umd.edu/bitstream/1903/70/2/dissertation.pdf; Hua Wang, Nlandu Mamingi, Benoit Laplante and Susmita Dasgupta, "Incomplete Enforcement of Pollution Regulation: Bargaining Power of Chinese Factories," Policy Research Working Paper 2756 (The World Bank, January 2002), 18–19, http://link.springer.com/article/10.1023%2FA%3A1022936506398; Benjamin van Rooij, "Implementation of Chinese Environmental Law: Regular Enforcement and Political Campaigns," *Development and Change*, Vol. 37, No. 1 (2006): 57, https://openaccess.leidenuniv.nl/bitstream/handle/1887/12533/artdandcFinalPubversion.pdf?sequence=2; Wang Canfa, "Chinese Environmental Law Enforcement: Current Deficiencies and Suggested Reforms," *Vermont Journal of Environmental Law*, Vol. 8, No. 2 (Spring 2007): 164; Richard L. Edmonds, *Patterns of China's Lost Harmony* (London: Routledge, 1994), 255–256. For additional research on implementation issues in China, see Zhang Yikai, "Towards Sustainable Development: Chinese Environmental Law Enforcement Mechanism Research," LLM thesis (University of Toronto, 2009), https://tspace.library.utoronto.ca/bitstream/1807/19011/6/Zhang_Yikai_200911_LLM_thesis.pdf; Jennifer Turner, "Authority Flowing Downwards? Local Government Entrepreneurship in the Chinese Water Sector," (Ann Arbor, MI: Indiana University, 1997); Michel Oksenberg and Elizabeth Economy, "China: Implementation under Economic Growth and Market Reform," in Harold K. Jacobson and Edith Brown Weiss eds., *Engaging Countries: Strengthening Compliance with International Environmental Accords* (Cambridge, MA: The MIT Press, 1998); Hon S. Chan, Koon-kwai Wong, and K. C. Cheung, "The Implementation Gap in Environmental Management in China: The Case of Guangzhou, Zhengzhou and Nanjing," *Public Administration Review*, Vol. 55, No. 4 (1995), 333–340; Barbara J. Sinkule, *Implementation of China's Three Synchronizations Policy: Case Studies of*

Wastewater Treatment Measures at New and Renovated Factories (Los Alamos, NM, 1994); Barbara J. Sinkule and Leonard Ortolano, *Implementing Environmental Policy in China* (Westport, CT: Praeger Publishers, 1995); Edward B. Vermeer, "Industrial Pollution in China and Remedial Policies," *The China Quarterly*, No. 156 (1998); Carlos Wing-Hung Lo, "Environmental Management By Law in China: The Guangzhou Experience," *The Journal of Contemporary China*, Vol. 6 (1994), 39–58.
2. At the 17th Party Congress in 2007, Chinese leaders began to talk about balancing economic development and environmental protection. Leaders included the concept of "ecological civilization" as one of the main components of an "all-around well-off society." Asian Development Bank, "Toward an Environmentally Sustainable Future: Country Environmental Analysis of the People's Republic of China," *Asian Development Bank*, August 2012, 92, http://www.adb.org/sites/default/files/pub/2012/toward-environmentally-sustainable-future-prc.pdf.
3. After MEP was established, authorities added 80 staff members. "Toward an Environmentally Sustainable Future: Country Environmental Analysis of the People's Republic of China," *Asian Development Bank*, August 2012, 95, http://www.adb.org/sites/default/files/pub/2012/toward-environmentally-sustainable-future-prc.pdf.
4. According to the ADB report, authorities created an Environmental Supervision Bureau in 2003 and expanded the number of local level supervision departments or divisions. As of 2012, there were 57,000 inspection staff working in 2,954 supervision departments or divisions. "Toward an Environmentally Sustainable Future: Country Environmental Analysis of the People's Republic of China," *Asian Development Bank*, August 2012, 96, http://www.adb.org/sites/default/files/pub/2012/toward-environmentally-sustainable-future-prc.pdf.
5. Asian Development Bank and the Ministry of Environmental Protection, "Market-based Policy Instruments for Water Pollution Control in China," Draft Final Report, submitted to the Asian Development Bank and the Ministry of Environmental Protection, May 2010, 6, http://www.google.com/url?sa=t&rct=j&q=%20operating%20permits%20polluting%20enterprises%20china&source=web&cd=8&cad=rja&ved=0CFcQFjAH&url=http%3A%2F%2Fhosted.comm100.com%2Fknowledgebase%2FDownload_ArticleAttachment.aspx%3Fid%3D100046%26siteid%3D88094&ei=zw25Ud6_JNWw4APp84HgBQ&usg=AFQjCNHfJcfvimYM0I9aGGO7HUTBY-iiwg&bvm=bv.47810305,d.dmg. Asian Development Bank, "Toward an Environmentally Sustainable Future: Country Environmental Analysis of the People's Republic of China," *Asian Development Bank*, August 2012, 112–118, http://www.adb.org/sites/default/files/pub/2012/toward-environmentally-sustainable-future-prc.pdf. According to a graph in the ADB report, authorities increased spending on pollution control from less than 1 percent of GDP in 2005, to 1.5 percent in 2008. China Green Development Index Report 2011, 308. According to the report, investment in environmental protection has been too low, in comparison with developed countries. Elizabeth Economy, "China's Environmental Future: The Power of the People," *McKinsey Quarterly*, June 2013, http://www.mckinsey.com/insights/asia-pacific/chinas_environmental_future_the_power_of_the_people. According

to Dr. Economy, China's investment in environmental protection has hovered around 1.3 percent of GDP.

6. "Environmental Protection Minister Emphasizes Strengthening Environmental Protection Accountability System Implement Local Government Responsibility" [Huanbao Buzhang Qiangdiao Jianquan Huanbao Wenze Zhidu Luoshi Difang Zhengfu Zeren], *China News Net*, March 2, 2009, http://www.china.com.cn/aboutchina/txt/2009-03/02/content_17355637.htm; State Council, "Opinion Regarding Strengthening Key Environmental Protection Work" [Guowuyuan guanyu jaiqiang huanjing baohu zhongdian gongzuo de yijian], October 17, 2011, item 1.3, http://www.zhb.gov.cn/ztbd/rdzl/hbgzyj/201110/t20111021_218646.htm. The Opinion calls on authorities to strengthen supervision and enforcement of environmental laws, as well as to establish an "enforcement responsibility system."

7. "Ecological civilization" is a complex concept that includes ideas of sustainable development; energy and resource conservation; conservation, protection, and restoration of the ecological environment; and environmental protection. Chinese Communist Party Constitution (2012), *Xinhua*, November 18, 2012, General Principles, English: http://news.xinhuanet.com/english/special/18cpcnc/2012-11/18/c_131982575.htm; Chinese: http://news.xinhuanet.com/18cpcnc/2012-11/18/c_113714762.htm. One rough translation of the paragraph on ecological civilization reads: "[t]he Party leads the people's construction of a socialist ecological civilization. Establish ideas of an ecological civilization that respect, comply with (accommodate?), and protect nature. Uphold the basic national policies of energy conservation and environmental protection. Uphold the guiding principle of the priority of conservation, protection, and restoration of the natural environment. Uphold growing production, prosperously living, and ecologically friendly ways of cultural development. In order to create a friendly environment for production and life, and to realize the Chinese nation's sustainable development, do the utmost to construct a conservationist and environmentally friendly society, and form patterns, industrial structures, production methods, and ways of living that conserve resources and protect the environment." Meng Si, "An Insight into the Green Vocabulary of the Chinese Communist Party," *Chinadialogue*, November 15, 2012, http://www.chinadialogue.net/article/show/single/en/5339. According to Meng Si, ecological civilization entered into policy documents first in 2007, when former Party Secretary and President Hu Jintao included the concept as being tied to the four basic goals of a "xiaokang society" (an all-around well-off society).

8. This chapter will not specifically address the question of whether environmental laws improve environmental quality. Nor will it specifically address the debate regarding the usefulness of deterrence over other types of incentives for compliance. It is a given that deterrence and enforcement does not always translate into compliance. In the China case, see Van Rooij, Benjamin and Lo, Carlos Wing-Hung, "Fragile Convergence: Understanding Variation in the Enforcement of China's Industrial Pollution Law," *Law & Policy*, Vol. 32, No. 1, (January 2010): 30. Available at SSRN: http://ssrn.com/abstract=1524521 or http://dx.doi.org/10.1111/j.1467-9930.2009.00309.x

9. Katarina Svatikova, "Economic Criteria for Criminalization: Optimizing Enforcement In Case of Environmental Violations," PhD thesis (2011), 108, http://repub.eur.nl/res/pub/23022/Marianne%20Breijer%20-%20Proefschrift%20Katarina%20Svatikova%5Blr%5D.pdf. For more information regarding the complexities of deterrence in the environmental sector, see "Compliance Theories," in Durwood Zaelke, Donald Kaniaru, and Eva Kružíková, eds., *Making Law Work: Environmental Compliance & Sustainable Development*, Vol. I, London: Cameron May Ltd, 2005, accessed through the International Network for Environmental Compliance and Enforcement, http://www.inece.org/mlw/Chapter 2_ComplianceTheories.pdf
10. Jon D. Silberman, "Does Environmental Deterrence Work? Evidence and Experience Say Yes, But We Need to Understand How and Why," *Environmental Law Reporter*, Vol. 30, (2000): 10523 http://elr.info/news-analysis/30/10523/does-environmental-deterrence-work-evidence-and-experience-say-yes-we-need (subscription required).
11. Michael Faure found that liability rules have a deterrent effect on potential polluters, pollution fees (taxes) of various types led to reductions in discharges and emissions. Michael Faure, *Effectiveness of Environmental Law: What Does the Evidence Tell Us?*; William and Mary, *Environmental Law & Policy Review*, Vol. 36, No. 2 (2012): 301, 311.
12. Hua Wang and David Wheeler, "Financial Incentives and Endogenous Enforcement in China's Pollution Levy System," *Journal of Environmental Economics and Management*, Vol. 49, No. 1 (2005): 174. Wang and Wheeler found financial penalties and self-reporting combined with few opportunities to contest the penalties acted as a deterrent on polluting behavior.
13. U.S. EPA, Poland Environment Ministry, and the Dutch Ministry, "The Basics for Compliance and Enforcement," *Principles of Environmental Enforcement*, (1992): 1, accessed through the International Network for Environmental Compliance and Enforcement, http://www.inece.org/princips/ch2.pdf; Gary S. Becker, "Crime and Punishment: An Economic Approach," *Journal of Political Economy*, Vol. 76, No. 2 (March–April 1968):169–217. Becker's early research found that specific regulatory instruments only work if sufficient enforcement and sanctions are in place.
14. Huanjing Jiufen Fangan Yu Chuli Shiwu Quanshu, *Environmental Disputes: A Practical Guide for Prevention and Management* [Zhongguo Yanshi Chubanshe] (China True Word Press, 1999), 20. Article 11 in the 1978 Constitution of the PRC reads "The state protects the environment and natural resources, and prevents and mitigates pollution and other public harms." The 1982 PRC Constitution, article 26 provides that the "state protects and improves the environmental and ecological conditions for life, and prevents pollution and other public harms." http://english.people.com.cn/constitution/constitution.html
15. National People's Congress, Environmental Protection Law of the People's Republic of China, see especially Articles 6, 8–11, 13–16, 24, 26–29, 35–36, 39–45. (These regulatory instruments will be discussed in more detail later.) http://www.china.org.cn/english/environment/34356.htm
16. Details regarding some of these mechanisms on this list are discussed in the subsequent sections of this chapter. The mechanisms listed in the two tables below were compiled based on the following resources: Michael

Palmer, "Environmental Regulation in the People's Republic of China: The Face of Domestic Law," *The China Quarterly*, Vol. 156 (1998); Zhao Ying, "A Survey of Environmental Law and Enforcement Authorities in China," *First International Conference on Environmental Compliance and Enforcement Proceedings*, Vol. 2, Chaing Mai, Thailand (April 21–26, 1996):3–10, http://www.inece.org/4thvol2/zhao.pdf; Hu Baolin, "Enforcement of Pollutant Discharge Fee in China," *Fourth International Conference on Environmental Compliance and Enforcement Proceedings*, Vol. 2, Chaing Mai, Thailand (April 21–26, 1996):3–3, http://www.inece.org/4thvol1/baolin.pdf; Barbara J. Sinkule, Leonard Ortolano, *Implementing Environmental Policy in China* (Westport, CT: Praeger Publishers, 1995); Barbara J. Sinkule, *Implementation of China's Three Synchronizations Policy: Case Studies of Wastewater Treatment Measures at New and Renovated Factories* (Los Alamos, NM 1994); Michael Palmer, "Environmental Regulation in the People's Republic of China: The Face of Domestic Law," *The China Quarterly*, Vol. 156 (1998); Carlos Wing-Hung Lo, "Environmental Management By Law in China: The Guangzhou Experience," *The Journal of Contemporary China*, Vol. 6 (1994); Moser, Adam J. and Yang, Tseming, "Environmental Tort Litigation in China," (November 1, 2011) *Environmental Law Reporter*, Vol. 41, 2011. Available at SSRN: http://ssrn.com/abstract=1821748; Wang, Hua, "Pollution Charges, Community Pressure, and Abatement Cost of Industrial Pollution in China," (May 2000) World Bank Policy Research Working Paper No. 2337. Available at SSRN: http://ssrn.com/abstract=630709; Dasgupta, Susmita, Laplante, Benoit, Mamingi, Nlandu and Wang, Hua, "Industrial Environmental Performance in China: The Impact of Inspections." (February 2000) World Bank Policy Research Working Paper No. 2285. Available at SSRN: http://ssrn.com/abstract=629142; Wang, Hua and Di, Wenhua, "The Determinants of Government Environmental Performance: An Empirical Analysis of Chinese Townships." (December 2002) World Bank Policy Research Working Paper No. 2937. Available at SSRN: http://ssrn.com/abstract=636298; Wang, Hua and Chen, Ming, "How the Chinese System of Charges and Subsidies Affects Pollution Control Efforts by China's Top Industrial Polluters." (October 1999) World Bank Policy Research Working Paper No. 2198. Available at SSRN: http://ssrn.com/abstract=623949; Wang, Hua, Mamingi, Nlandu, Dasgupta, Susmita and Laplante, Benoit, "Incomplete Enforcement of Pollution Regulation: Bargaining Power of Chinese Factories." (January 18, 2002) World Bank Policy Research Working Paper No. 2756. Available at SSRN: http://ssrn.com/abstract=634469; The Environmental Protection Law of the People's Republic of China, issued 1989; Organization for Economic Cooperation and Development (OECD). 2006. *Environmental Compliance and Enforcement in China: An Assessment of Current Practices and Steps Forward: An Assessment of Current Practices and Ways Forward*, draft report presented at the second meeting of the Asian Environment Compliance and Enforcement Network, December 4–6, http://www.oecd.org/environment/outreach/3786751l.pdf; *OECD Environmental Performance Reviews: China*, OECD. For more information on China's total load control and pollution discharge permit systems, see Li Zhiping, "The Challenges of China's Discharge Permit System and Effective Solutions," *Temple Journal*

of Science, Technology, and Environmental Law, Vol. XXIV, No. 1 (Spring 2005):375–395. For additional information on some of the early pilot policies to control water pollution, see "Market-based Policy Instruments for Water Pollution Control in China," Draft Final Report, submitted to the Asian Development Bank and the Ministry of Environmental Protection, May 2010, http://www.google.com/url?sa=t&rct=j&q=%20operating %20permits%20polluting%20enterprises%20china&source= web&cd=8&cad=rja&ved=0CFcQFjAH&url=http%3A%2F%2Fhosted. comm100.com%2Fknowledgebase%2FDownload_ArticleAttachment.aspx% 3Fid%3D100046%26siteid%3D88094&ei=zw25Ud6_JNWw4APp84 HgBQ&usg=AFQjCNHfJcfvimYM0I9aGGO7HUTBY-iiwg&bvm=bv. 47810305,d.dmg.
17. To understand which rules and documents issued by different official sources have coercive power, see State Council, "Measures for Handling Official Documents in State Administrative Organs" [Guojia xingzheng jiguan gongwen chuli banfa], issued August 24, 2000, art. 9.
18. The law provides for criminal sanctions (Article 57), but does not provide specific standards for the application of the Criminal Law. "Circular Economy Promotion Law of the People's Republic of China" [Zhongguo Gonghe Renmin Gongheguo Xunhuan Jingji Cujinfa], NPC Standing Committee, passed August 29, 2008, effective January 1, 2009, Section 6. http://zfs.mep.gov.cn/fl/200809/ t20080901_128001.htm (Chinese) http://www.amcham-shanghai.org/ NR/rdonlyres/4447E575-58FD-4D8E-BB0F-65B920770DF7/7987/ CircularEconomyLawEnglish.pdf (English)
19. Each of the following plans or measures has articles or chapters on the legal liabilities for violations. State Council, "National Plan for Environmental Emergency Response" [Guojia Tufa Huanjing shijian Yingji Yu'an], issued, January 1, 2006, http://zfs.mep.gov.cn/fg/xzhg/ 200611/t20061121_96328.htm. Ministry of Environmental Protection, "Interim Measure for Environmental Emergency Response Plan Management" [Tufa Huanjing Shijian Yingji Yu'an Guanli Zhanxing Banfa], issued September 28, 2010, http://www.mep.gov.cn/gkml/hbb/bwj/ 201010/t20101009_195330.htm. Ministry of Environmental Protection, "Sudden Environmental Incident Information Reporting Measure" [Tufa Huanjing Shijian Xixi Baogao Banfa], issued March 24, 2011, Article 15.
20. Ministry of Environmental Protection, "Measures for On-the-spot Inspections of Pollution Source Automated Monitoring Facilities" [Wuranyuan Zidong Jiankong Shishi Xianchang Jiandu Jiancha Banfa], issued December 30, 2011, http://www.mep.gov.cn/gkml/hbb/bl/201202/ t20120213_223413.htm. Articles 17–26 stipulate legal liabilities that refer to Articles in the Water Pollution Prevention and Control Law and the Air Pollution Prevention and Control Law, which include specific monetary penalties.
21. Criminal Law of the People's Republic of China (2011 revision), Article 338, http://www.szxingshi.com/95w9.html. See the following National People's Congress web site for links to all the previous amendments and the 1997 version of the law, as well as numerous supporting documents: http://law-lib.com/law/law_view.asp?id=327&page=7

22. For example, both versions of the Criminal Law stipulate in Article 338 that violators of the law will be sentenced to less than three years in prison or criminal detention and fined or required to pay a one-time penalty. In cases where the consequences are extremely severe will be sentenced to more than three years but less than seven years, and issued a fine. Criminal Law of the People's Republic of China (2011 revision), Article 338, http://www.szxingshi.com/95w9.html. Similar punishments are stipulated for many of the other environmental crimes detailed in Articles 38 through 46.
23. The table linked from the site below clearly shows the differences between the 2006 and 2013 versions of the interpretation. "Supreme People's Court and Supreme People's Procuratorate Interpretation of Certain Issues Related to Laws Applicable in Criminal Cases of Environmental Pollution" [Zuigao renmin fayuan, zuigao renmin jianchayuan guanyu banli huanjing wuran xingshi anjian shiyong falu ruogan wenti de jieshi], June 17, 2013, http://www.chinacourt.org/law/detail/2013/06/id/146211.shtml
24. "Congressional-Executive Commission on China, 2012 Annual Report, Section 2—The Environment," 115. http://www.gpo.gov/fdsys/pkg/CHRG-112shrg76190/pdf/CHRG-112shrg76190.pdf (based on the following reports: "At the Start Standing Is Not Obvious, Probably Few Public Interest Lawsuits for the Time Being" [Qidong zhuti buminglang gongyi susong huo zhanshi bu duo], *Legal Daily*, September 4, 2012; "Expert Opinion: Public Interest Lawsuit Procedural System Still Not Independent" [Zhuanjia guandian: gongyi susong chengxu zhidu youdai duli], *Legal Daily*, September 4, 2012; Chen Liping, "Wang Shengming: Standing in Public Interest Lawsuits Could Be Clarified By Relevant Laws" [Wang shengming: gongyi susong zhuti ke you xiangguan falu mingque], *Legal Daily*, September 4, 2012.)
25. For more information about these cases, see the following reports. Yan Zhijiang, "All-China Environment Federation in Guiyang Wins Environment Public Interest Litigation Case" [Zhonghua huanbao lianhehui guiyang daying huanjing gongyi susong'an], *Legal Daily*, January 4, 2011; Qie Jianrong, "Chromium Slag Pollution Case for 10 Million in Damages Already Formally Accepted by Court" [Yin gezha wuran suopei qianwan an yi zhengshi lian], *Legal Daily*, October 20, 2011; "Difficulties with Environmental Public Interest Suits: Hard to Obtain Evidence, Assessment Costs High" [Huanjing gongyi susong zhi kun: quzheng nan pinggu feiyong gao], *China Weekly*, reprinted in Sina, April 16, 2012;"Talks Begin in Landmark NGO Environment Case," *China Daily*, May 24, 2012. Chang Cheng, "China's Shifting Public Space," *Chinadialogue*, June 21, 2012, http://www.chinadialogue.net/article/show/single/en/5000-China-s-shifting-public-space. According to Chang Cheng, all of the public interest cases in court between 2007 and 2010 were brought by the All-China Environmental Federation, which is a group affiliated with the Ministry of Environmental Protection. http://www.chinadialogue.net/article/show/single/en/5000-China-s-shifting-public-space
26. "Eight Cases That Mattered," *Chinadialogue*, July 26, 2011, http://www.chinadialogue.net/article/show/single/en/4429-Eight-cases-that-mattered. The first case in which a procuratorate filed a public interest suit occurred in Guangdong and involved pollution from a textile factory that discharged

pollution into the Shiliugang River. The Haizhu Procuratorate filed a suit against the boss of the textile plant in the Guangzhou Maritime Court, which ruled the boss was liable. The court ordered him to pay over 115,000 Yuan in compensation. The procuratorate argued that it had standing to bring the case because the river was a natural resource and the state was legally responsible for its "legal supervision" (based on Article 3 of the Water Law and Article 73 of the General Principals of Civil Law). According to this article, there have been other cases in Guangdong where the procuratorate filed a court case, although there have not been cases in other provinces.

27. "China's 10 Most Influential Court Cases in 2012" [2012 nian zhongguo shida yingxiangxing susong], *Southern Weekend*, January 10, 2013; "Mediation Fails in Yunnan Chromium Slag Public Interest Case, Defendant Unilaterally Backs Out" [Yunnan gezha gongyi susong an tiaojie shibai yin beigao danfang mianfanhui], *Justice Net*, April 19, 2013. According to the above articles, the case, as of June 2013 had not gone to court. The parties involved had come to an agreement but the defendant pulled out of the agreement at the last minute. The facility had agreed to accept legal and financial liability, promised to stop the polluting behavior as well as to restore the environment. The defendant also agreed to third-party oversight of their implementation of the agreement. For background information, see "Congressional-Executive Commission on China, 2012 Annual Report, Section 2—The Environment," 115. http://www.gpo.gov/fdsys/pkg/CHRG-112shrg76190/pdf/CHRG-112shrg76190.pdf, based on the following sources: Qie Jianrong, "Chromium Slag Pollution Case for 10 Million in Damages Already Formally Accepted by Court" [Yin gezha wuran suopei qianwan an yi zhengshi lian], *Legal Daily*, October 20, 2011; "Difficulties with Environmental Public Interest Suits: Hard to Obtain Evidence, Assessment Costs High" [Huanjing gongyi susong zhi kun: quzheng nan pinggu feiyong gao], *China Weekly*, via Sina, April 16, 2012; In May, the environmental tribunal under the Qujing Intermediate People's Court presided over pretrial negotiations and the court reportedly had two meetings about the case in July and August. For more information, see Cao Yin and Guo Anfei, "Talks Begin in Landmark NGO Environment Case," *China Daily*, May 24, 2012; Friends of Nature, "Green Protests on the Rise in China," August 14, 2012.

28. "Congressional-Executive Commission on China, 2012 Annual Report, Section 2—The Environment," 115. http://www.gpo.gov/fdsys/pkg/CHRG-112shrg76190/pdf/CHRG-112shrg76190.pdf, based on the following sources: Yan Zhijiang, "All-China Environment Federation in Guiyang Wins Environment Public Interest Litigation Case" [Zhonghua huanbao lianhehui guiyang daying huanjing gongyi susong'an], *Legal Daily*, January 4, 2011; "First Local Environmental Public Interest Litigation Case Trial Opened December 30" [Bentu huanjing gongyi susong diyi an 12 yue 30 ri kaiting shenli], *Guiyang News Net*, reprinted in Guiyang Public Environmental Education Center, December 31, 2010; "Guizhou First Non-Governmental Organization Filed Environmental Public Interest Lawsuit Enters Judicial Process" [Guizhou shouli minjian huanbao zuzhi tiqi de huanjing gongyi susong jinru sifa chengxu], Guiyang Public Environmental Education Center, November 23, 2010.

29. "Eight Cases That Mattered," *Chinadialogue*, July 26, 2011. http://www.chinadialogue.net/article/show/single/en/4727-China-s-courts-fail-the-environment-
30. The Civil Procedure Law of the People's Republic of China (2012 revisions), Articles 63, 76–79. Article 76 stipulates that the court designate an expert witness if both parties cannot agree on the credentials of an expert witness proposed by one party. Article 78 stipulates that if a party objects to an expert's opinion, that party may apply to the court to have the expert witness testify in court. Article 79 stipulates that a party may apply to the court for permission to call on individuals with specialized knowledge to testify. http://www.npc.gov.cn/npc/xinwen/2012-09/01/content_1735841.htm (2012 version). The previous version, the Civil Procedure Law of the People's Republic of China (2007), Article 72 stipulates that the court may call for an "expert conclusion" from relevant government departments, but did not give the parties the right to apply for permission to call on their own expert witnesses. http://law.npc.gov.cn:87/page/secondbrw.cbs?rid=1&order=10&result=C%3A%5CWINDOWS%5CTEMP%5CB283153%2Etmp&page=allindex&f=&field=&transword=++%D6%D0%BB%AA%C8%CB%C3%F1%B9%B2%BA%CD%B9%FA%C3%F1%CA%C2%CB%DF%CB%CF%B7%A8&dkall=1&OpenCondition=FULLTEXT%3D%27%28%23%CA%B1%D0%A7%D0%D4%3D%2A%29+AND++%28%23%B0%E4%B2%BC%CA%B1%BC%E4%3E%3D19891209+AND+%23%B0%E4%B2%BC%CA%B1%BC%E4%3C%3D20110301%29+AND++%28%D6%D0%BB%AA%C8%CB%C3%F1%B9%B2%BA%CD%B9%FA%C3%F1%CA%C2%CB%DF%CB%CF%B7%A8%29%27
31. "Congressional-Executive Commission on China 2012 Annual Report," October 10, 2012, 115, http://www.gpo.gov/fdsys/pkg/CHRG-112shrg76190/pdf/CHRG-112shrg76190.pdf based on the following reports: Wang Hairong, "Thwarting Dirty Migration," *Beijing Review*, No. 6, February 9, 2012. An official at the All-China Environment Federation asserted that the courts reject most public interest lawsuits because of legal standing problems. Authorities were reportedly also concerned about an onslaught of environmental litigation cases. Xia Jun, "China's Courts Fail the Environment," *Chinadialogue*, January16, 2012. Additional reasons why courts do not accept cases include government interference in the decision, challenges in assessing environmental damages, and concern for maintaining "social stability."
32. According to research regarding trends in enforcement of environmental pollution violations, there was a 208 percent increase in the average amount of a fine per case between 2001 and 2006 and the number of fines levied also increased over that period of time. Van Rooij, Benjamin and Lo, Carlos Wing-Hung, "Fragile Convergence: Understanding Variation in the Enforcement of China's Industrial Pollution Law," *Law & Policy*, Vol. 32, No 1 (January 2010):17–18. Available at SSRN: http://ssrn.com/abstract=1524521 or http://dx.doi.org/10.1111/j.1467-9930.2009.00309.x
33. For more details, see Squire Sanders, "China Introduces New Rules for Managing Hazardous Chemicals," 2011, accessed February 15, 2013, http://www.squiresanders.com/files/Publication/25ba90a3-f867-4df3-8fa7-bb5e75822766/Presentation/PublicationAttachment/844ccf91

-2330-42fc-acb5-bd837b12cb78/China_Alert_-_China_Introduces_New_ Rules_for_Managing_Hazardous_Chemicals_by_Squire_Sanders.pdf; "Regulations on the Safe Management of Hazardous Chemicals" [Weixian Huaxueping Anquan Guanli Tiaoli], *State Council*, issued January 26, 2002, amended February 16, 2011. http://bio.cbi.pku.edu.cn/equipment/4.doc (2002); http://zfs.mep.gov.cn/fg/xzhg/201103/t20110314_206673.htm (2011 revisions). English version of the 2011 regulations – http://www.cirs-reach.com/China_Chemical_Regulation/Regulations_on_safe_management_on_hazardous_chemicals_China_2011.pdf. For more information on the hazardous chemical management system in China, see http://www.env.go.jp/chemi/reach/second/gao.pdf

34. Environmental Protection Administrative Punishment Measure (1999) [Huanjing baohu xingzheng chufa banfa], Issued, August 6, 1999, art. 17, http://english.mep.gov.cn/Policies_Regulations/regulations/Environmental_Enforcement/200712/t20071203_113699.htm (English).

35. Article 43 of the 1999 measure clearly stipulates the additional fine of "3 percent of the amount of the delinquent fine per day." The 2009 version of the law does not contain this stipulation, decreasing the deterrence effect of the measure. Ministry of Environmental Protection, "Environmental Administrative Penalty Measure" [Huanjing Zingzheng Chufa Banfa], passed December 30, 2009, effective March 1, 2010. http://www.mep.gov.cn/gkml/hbb/bl/201002/t20100201_185230.htm

36. For full information, see "Environmental Administrative Penalty Measure" [Huanjing Zingzheng Chufa Banfa], passed December 30, 2009, effective March 1, 2010. http://www.mep.gov.cn/gkml/hbb/bl/201002/t20100201_185230.htm

37. "Implementing Regulations for the Water Pollution Law of the People's Republic of China" [Zhonghua Renmin Gongheguo Shui Wuran Fangzhifa Xize], Order of the State Council, March 20, 2000, Article 38.2. http://www.gov.cn/gongbao/content/2000/content_60099.htm; "Water Pollution Law of the People's Republic of China" [Zhonghua Renmin Gongheguo Shui Wuran Fangzhifa], Standing Committee of the National People's Congress, passed May 11, 1984, amended May 15, 1996, amended February 28, 2008, Article 70, http://zfs.mep.gov.cn/fl/200802/t20080229_118802.htm

38. Water Pollution Law Implementing Regulations, Article 39.1.

39. Water Pollution Law, Article 76.2.

40. For the full list of environmental targets, see National People's Congress, "Outline of the 12th Five-Year Economic and Social Development Plan" [Guojia huanjing baohu "shieyiu" guihua de tongzhi], issued March 2011, Part I, Section 3, para 5, among others, http://www.npc.gov.cn/wxzl/gongbao/2011-08/16/content_1665636.htm.

41. For comparison, see the National People's Congress, "Outline of the 11th Five-Year Economic and Social Development Plan" [Guojia huanjing baohu "shiyiwu" guihua de tongzhi], issued March 16, 2006, Part I, Section 3, para. 4, http://news.xinhuanet.com/misc/2006-03/16/content_4309517_1.htm

42. Authorities issue annual reports tracking the progress toward goals in the five-year plans, see Report on the Implementation of the 2010 Plan for National Economic and Social Development and on the 2011 Draft Plan

for National Economic and Social Development, Xinhua, March 17, 2011, item 4, http://news.xinhuanet.com/english2010/china/2011-03/17/c_13783842_5.htm

43. "Ministry of Environmental Protection, Measures for the Ex-post Supervision of Environmental Administrative Enforcement" [Huanjing Xingzheng zhifa ho ducha banfa], issued, November 5, 2010. Article 6 indicates when officials should undertake post hoc supervision activities and Articles 9–10 indicate steps officials may take when a party refuses to comply with specified administrative penalties, http://www.mep.gov.cn/gkml/hbb/bl/201012/t20101217_198805.htm.

44. "Administrative Coercion Law of the People's Republic of China" [Zhonghua Renmin Gongheguo Xingzheng Qiangzhi Fa], passed by the National People's Congress Standing Committee, June 30, 2011, effective January 1, 2012. http://zfs.mep.gov.cn/fl/201107/t20110701_214304.htm. Previously, the Water Pollution Prevention and Control Law (2008 revisions) included similar stipulations. "Water Pollution Prevention and Control Law of the People's Republic of China" [Zhonghua Renmin Gongheguo Shui Wuran Fangzhifa] (2008 revisions), Standing Committee of the National People's Congress, passed February 28, 2008, effective June 1, 2008. http://zfs.mep.gov.cn/fl/200802/t20080229_118802.htm.

45. "Administrative Coercion Law of the People's Republic of China" [Zhonghua Renmin Gongheguo Xingzheng Qiangzhi Fa], passed by the National People's Congress Standing Committee, June 30, 2011, effective January 1, 2012. http://zfs.mep.gov.cn/fl/201107/t20110701_214304.htm

46. National People's Congress, The General Principles of Civil Law of the People's Republic of China, issued April 12, 1986. Article 124 states "any person who pollutes the environment and causes damage to others in violation of state provisions for environmental protection and the prevention of pollution shall bear civil liability in accordance with the law," http://www.law-lib.com/law/law_view.asp?id=3633; National People's Congress, Environmental Protection Law of the People's Republic of China, issued December 26, 1989. Article 41 states "a unit that has caused an environmental pollution hazard shall have the obligation to eliminate it and make compensation to the unit or individual that suffered direct losses." http://www.china.org.cn/english/environment/34356.htm. For a more in-depth discussion of the complexities of the legal framework for pollution compensation claims, including debates about no-fault liability and contradictions among various legal statutes, and rules regarding class action suits, types of remedies, and the statute of limitations, see Alex Wang "The Role of Law in Environmental Protection in China: Recent Developments," *Vermont Journal of Environmental Law*, Vol. 8 (2006–2007), http://www.vjel.org/journal/VJEL10057.html.

47. The remarks regarding the Environmental Protection Law were taken from the following source: Congressional-Executive Commission on China, "Environmental Protection Law Draft Revisions: Authorities Remove Language Regarding Strengthening Public Participation, Accountability, and Transparency," March 13, 2012, http://www.cecc.gov/pages/virtualAcad/index.phpd?showsingle=169174

48. National People's Congress, "Draft Laws for Public Comment," accessed January 1, 2013, http://www.npc.gov.cn/npc/flcazqyj/node_8176.htm.
49. "Environment Law to Take Back Seat at NPC Meeting," *South China Morning Post*, February 26, 2013, http://www.scmp.com/news/china/article/1158565/bold-lawmaking-take-back-seat-npc-meeting; Ministry of Environmental Protection, "Four Major Problems with the Environmental Protection Law Amendment Draft," November 1, 2012, http://news.sina.com.cn/c/2012-11-01/103825487296.shtml
50. For a more in-depth discussion of the drafting process and the provisions authorities removed from the draft for public comment, see Congressional-Executive Commission on China, "Environmental Protection Law Draft Revisions: Authorities Remove Language Regarding Strengthening Public Participation, Accountability, and Transparency," March 13, 2012, http://www.cecc.gov/pages/virtualAcad/index.phpd?showsingle=169174
51. Wang Shekun, "Regulations on Environmental Impact Assessment for Plans: Its Breakthroughts and Limitations," *Friends of Nature*, accessed, February 23, 2013: 6, http://www.fon.org.cn/uploads/attachment/83521361859986.pdf
52. The sources for these number are: "2006 Environmental Specialized Operations 28,000 Environmental Violation Suits Filed of Cases Investigated" [2006 Nian Huanbao Zhuanxiang Xingdong Lian Chachu Huanjing Weifa 2.8 Wan Jian], *China Net*, May 5, 2007, http://www.china.com.cn/environment/txt/2007-05/05/content_8209241.htm; "Ministry of Environmental Protection, 2007 Report on China's Environmental Conditions" [2007 Nian Zhongguo Huanjing Zhuangkuang Gongbao], *Special Topics*, November 17, 2008, http://jcs.mep.gov.cn/hjzl/zkgb/2007zkgb/200811/t20081117_131285.htm; Ministry of Environmental Protection, "2008 Report on China's Environmental Conditions" [2008 Nian Zhongguo Huanjing Zhuangkuang Gongbao], *Special Topics*, June 9, 2009, http://jcs.mep.gov.cn/hjzl/zkgb/2008zkgb/200906/t20090609_152542.htm; "In 09 Investigate and Handle 2,183 Enterprise Environmental Violations Outlaw and Close 231" [09 Chachu Huanjing Weifa Qiye 2183 Jia Qudi Guanbi 231 Jia], *China News*, June 3, 2010, http://www.china.com.cn/news/2010-06/03/content_20177034.htm; Ministry of Environmental Protection, "2010 Report on China's Environmental Conditions" [2010 Nian Zhongguo Huanjing Zhuangkuang Gongbao], *Special Topics*, June 3, 2011, http://jcs.mep.gov.cn/hjzl/zkgb/2010zkgb/201106/t20110602_211559.htm; "Ministry of Environmental Protection, 2011 Report on China's Environmental Conditions" [2011 Nian Zhongguo Huanjing Zhuangkuang Gongbao], *Special Topics*, June 6, 2012, http://jcs.mep.gov.cn/hjzl/zkgb/2011zkgb/201206/t20120606_231062.htm; Ministry of Environmental Protection, "2012 Report on China's Environmental Conditions" [2012 Zhongguo Huanjing Zhuangkuang Gongbao], 2, May 28, 2013, http://www.mep.gov.cn/gkml/hbb/qt/201306/W020130606578292022739.pdf.
53. The State Council issued the decision to examine environmental violations over a five-year period. "China Turns the Corner in Reducing Emissions" [Zhongguo Jianpai Yinglai Guaidian], *China Electricity Information Net*, via National Electricity Supervision Committee, July 23, 2008, http://njb.serc.gov.cn/news/2008-7/200872485425.htm.

54. For more information about the trouble citizens encounter when they try to file environmental lawsuits, see "Congressional-Executive Commission on China 2012 Annual Report," October 10, 2012, 115, http://www.gpo.gov/fdsys/pkg/CHRG-112shrg76190/pdf/CHRG-112shrg76190.pdf; Xia Jun, "China's Courts Fail the Environment," *Chinadialogue*, January 16, 2012; "Congressional-Executive Commission on China 2011 Annual Report," October 10, 2011, 141, http://www.cecc.gov/pages/annualRpt/annualRpt11/AR2011final.pdf.
55. Xia Jun, "China's Courts Fail the Environment," *Chinadialogue*, January 16, 2012, http://www.chinadialogue.net/article/show/single/en/4727-China-s-courts-fail-the-environment-
56. For example, see the preface of the Circular on Enhancing the Environmental Monitoring over Export Enterprises [Guanyu Jiaqiang Chukou Qiye Huanjing Jianguan de Tongzhi], issued October 8, 2007, http://english.mofcom.gov.cn/aarticle/policyrelease/announcement/200803/20080305438742.html (English), http://english.mofcom.gov.cn/aarticle/subject/cv/lanmuc/200803/20080305438749.html (Chinese). See also the Ministry of Environmental Protection, "Adjusting and Improving Economic Policies," April 11, 2006, http://gcs.mep.gov.cn/hjgh/sywgh/dsywgh/200604/t20060411_75669.htm
57. Motoko Aizawa, "China's Green Credit Policy: Building Sustainability in the Financial Sector," World Resources Institute, June 8, 2011, http://powerpoints.wri.org/gei_book_launch/aizawa_chinas_green_credit_policy_2011-06.pdf.
58. Note that these measures only apply to a small subset of Chinese companies. Ministry of Environmental Protection Circular on Environmental Protection Verification of Highly Polluting Industries to Further Standardize the Production and Management of Companies Applying for Public Listing or Refinancing [Guanyu Jinyibu Guifan Zong Wuran Hangye Shengchan Jingying Congsi Shengqing Shangshi Huo Zai Rongzi Huanjing Baohu Hecha Gongzuo de Tongzhi], issued August 13, 2007, http://www.mep.gov.cn/info/gw/huanban/200708/t20070816_107999.htm
59. There are two relevant measures; see "Guiding Opinions Regarding Strengthening Environmental Protection Supervision and Management of Listing Companies" [Guanyu Jiaqiang Shangshi Congsi Huanjing Baohu Jiandu Guanli Gongzuo de Zhidao Yijian], issued February 22, 2008, http://www.mep.gov.cn/info/gw/huanfa/200802/t20080225_118650.htm; "Management of Listing Companies and the Public Company Environmental Protection Examination Industry Classification Management Directory" [Shangshi Gongsi Huanbao Hecha Hangye Fenlei Guanli Minglu], issued June 24, 2008, http://www.zhb.gov.cn/info/bgw/bbgth/200807/t20080707_125138.htm. For more information on the green securities policies and up-to-date information, see Hua Wang, Assessing the Green Securities Policy in China: An Examination of Environmental Disclosure by Chinese Listed Companies in Highly Polluting Industries, MPP Essay, Oregon State University, June 20, 2012, http://ir.library.oregonstate.edu/xmlui/bitstream/handle/1957/32934/Wang_MPP_Essay.pdf?sequence=1.
60. Note that there is not a requirement to make the names public. Ministry of Commerce and the State Environmental Protection Administration, "Circular on Enhancing the Environmental Monitoring Over Export

Enterprises" [Guanyu Jiaqiang Chukou Qiye Huanjing Jianguan de Tongzhi], issued October 8, 2007, http://english.mofcom.gov.cn/aarticle/policyrelease/announcement/200803/20080305438742.html (English), http://english.mofcom.gov.cn/aarticle/subject/cv/lanmuc/200803/20080305438749.html (Chinese).

61. "China Ushers in 'Green Insurance System' to Curb Pollution," *Xinhua via Chinaview.cn*, February 18, 2008, http://news.xinhuanet.com/english/2008-02/18/content_547625416.htm

62. Ministry of Environmental Protection and Insurance Regulatory Commission, "Guiding Opinions regarding Launching Pilot Work on Compulsory Environmental Pollution Liability Insurance" [Guanyu Kaizhan Huanjing Wuran Qiangzhi Xeren Baoxian Shidian Gongzuo de Zhidao Yijian], January 21, 2013.

63. "Compulsory Insurance for Environment Risks," *Shanghai Daily*, February 22, 2013.

64. State-Owned Assets Supervision and Administration Commission, "Environmentally Responsible Corporate Conduct in China," *OECD Investment Policy Reviews: China 2008*, OECD, 2008, 262, http://www.oecd.org/daf/inv/investmentfordevelopment/41792807.pdf. For more information on corporate social responsibility in China, see Simon Zadek, Maya Forstater, and Kelly Yu, "Corporate Responsibility and Sustainable Economic Development in China: Implications for Business," *U.S. Chamber of Commerce, National Chamber Foundation*, March 2012, http://www.uschamber.com/sites/default/files/international/files/17296_China%20Corp%20Social%20Responsibility_Opt.pdf

65. For example, efforts to prohibit companies from listing on the Chinese stock market if they have an unfavorable environmental performance applies only to companies large enough to be listed on the market. Anna Brettell, "The 'Scientific Outlook on Development' and the Future of Environmental Protection," *CSIS Freeman Report* (November 2007).

66. Xu Nan, "Chinese Banks under Almost Negligible Pressure to Protect the Environment," *Chinadialogue*, March 21, 2013, http://www.chinadialogue.net/article/show/single/en/5812-Chinese-banks-under-almost-negligible-pressure-to-protect-the-environment.

67. State Environmental Protection Administration, "Provisional Measures on Open Environmental Information" [Huanjin Xinxi Gongkai Banfa], issued April 11, 2007, effective May 1, 2008, via the Congressional-Executive Commission on China, http://www.cecc.gov/pages/virtualAcad/index.phpd?showsingle=95188. Other ministries that also have some authority over environmental affairs also passed their own versions of the OGI measures.

68. These catalogs are available on the MEP website. Ministry of Environmental Protection, "Ministry of Environmental Protection Information Guide" [Huanjing Baohubu Xinxi Gongkai Zhinan], 2008, http://www.zhb.gov.cn/gkml/hbb/qt/200910/t20091030_180595.htm; and Ministry of Environmental Protection, "Open Information Catalogue (First Group)" [Huanjing Baohubu Xinxi Gongkai Mulu (Diyipi)], 2008, http://www.mep.gov.cn/gkml/hbb/bgg/200910/W020080507275848932327.pdf.

69. For more information, see Congressional-Executive Commission on China, "China Commits to 'Open Government Information' Effective

May 1, 2008," May 2, 2008, http://www.cecc.gov/pages/virtualAcad/index.phpd?showsingle=105351; Congressional-Executive Commission on China, "SEPA Issues Measures on Open Environmental Information," January 31, 2008, http://www.cecc.gov/pages/virtualAcad/index.phpd?showsingle=101942.
70. For roundup information regarding implementation of the measures, see "Congressional-Executive Commission on China, 2009 Annual Report," 198–199, http://www.cecc.gov/pages/annualRpt/annualRpt09/CECCannRpt2009.pdf; "Congressional-Executive Commission on China, 2010 Annual Report," 154–156, http://www.cecc.gov/pages/annualRpt/annualRpt10/CECCannRpt2010.pdf; "Congressional-Executive Commission on China, 2011 Annual Report," 145–146, http://www.cecc.gov/pages/annualRpt/annualRpt11/AR2011final.pdf; "Congressional-Executive Commission on China, 2012 Annual Report," 117–118, http://www.gpo.gov/fdsys/pkg/CHRG-112shrg 76190/pdf/CHRG-112shrg 76190.pdf.
71. For more information, see Congressional-Executive Commission on China, "China Commits to 'Open Government Information' Effective May 1, 2008," May 2, 2008, http://www.cecc.gov/pages/virtualAcad/index.phpd?showsingle=105351; Congressional-Executive Commission on China, "SEPA Issues Measures on Open Environmental Information," January 31, 2008, http://www.cecc.gov/pages/virtualAcad/index.phpd?showsingle=101942.
72. For annual assessments of the progress and variation in implementation of the measures, see Natural Resources Defense Council and Institute for Public and Environmental Affairs, "Breaking the Ice on Open Environmental Information" [Huanjing xinxi gongkai jiannan pobing], June 2010, http://www.ipe.org.cn/uploadfiles/2009-08/1251342198040.pdf; Natural Resources Defense Council and Institute of Public & Environmental Affairs, "Environmental Open Information: Between Advance & Retreat—The 2009–2010 Pollution Information Transparency Index (PITI) Second Annual Assessment of Environmental Transparency in113 Chinese Cities," December 28, 2010, http://www.ipe.org.cn/Upload/Report-PITI-09-10-EN.pdf; Natural Resources Defense Council and Institute of Public and Environmental Affairs (IPE), "Open Environmental Information: Taking Stock," January 16, 2012, http://www.ipe.org.cn/Upload/Report-PITI-2011-EN.pdf.
73. Human Rights in China, "State Secrets: China's Legal Labyrinth," June 12, 2007, 174, http://www.hrichina.org/sites/default/files/oldsite/PDFs/State-Secrets-Report/HRIC_StateSecrets-Report.pdf. There may be additional regulations or rules that have not been made public.
74. Ministry of Environmental Protection, "Circular on Taking Steps to Strengthen Environmental Open Government Information Work" [Guanyu Jinyibu Jiaqiang Huanjing Baohu Xinxi Gongkai Gongzuo de Tongzhi], issued October 30, 2012, http://www.mep.gov.cn/gkml/hbb/bgt/201210/t20121031_240771.htm. The document is only a circular, so it has limited coercive power.
75. Congressional-Executive Commission on China, "Authorities Issue Circular to Promote Environmental Information Disclosure," February 4, 2013, http://www.cecc.gov/pages/virtualAcad/index.phpd?showsingle=

185443. For example, one shortcoming is that the circular is only applicable to a limited number of institutions.
76. Ministry of Environmental Protection, "Measures for the Ex-post Supervision of Environmental Administrative Enforcement" [Huanjing Xingzheng zhifa hou ducha banfa], issued, November 5, 2010, Article 11, http://www.mep.gov.cn/gkml/hbb/bl/201012/t20101217_198805.htm
77. "Environmental Administrative Penalty Measure" [Huanjing Zingzheng Chufa Banfa], passed December 30, 2009, effective March 1, 2010, Art. 72. http://www.mep.gov.cn/gkml/hbb/bl/201002/t20100201_185230.htm
78. State Environmental Protection Administration, "Provisional Measures on Public Participation in Environmental Impact Assessments" [Huanjin Yingxiang pingjia gonggong canyu zhanshi banfa], issued February 22, 2006, effective March 18, 2006, http://www.gov.cn/jrzg/2006-02/22/content_207093.htm.
79. For information on the implementation of these measures, see Jesse Moorman and Zhang Ge, "Promoting and Strengthening Public Participation in China's Environmental Impact Assessment Process: Comparing China's EIA Law and U.S. NEPA," *Vermont Journal of Environmental Law*, Vol. 8 (2006–2007):282–308; Zhang Jingjing, "The Plight of the Public," *Chinadialogue*, July 19, 2007, http://www.chinadialogue.net/article/show/single/en/4414-The-plight-of-the-public-1-, Zhao Yuhong, "Public Participation in China's EIA Regime: Rhetoric or Reality?" *Journal of Environmental Law*, Vol. 22, No. 1 (2010), http://jel.oxfordjournals.org/content/22/1/89.abstract.
80. Ministry of Environmental Protection Environmental Standards Research Institute, "Letter Regarding Solicitation of Comments on the National Environmental Protection Standard 'Technical Guidelines for Public Participation in Environmental Impact Assessments' (Draft for Comment)" [Guanyu Zhengqiu Guojia Huanjing Baohu Biaojun "Huanjin Yingxiang Pingjia Jishu Daoze Gongzhong Canyu" (Zhengqiu Yijian Gao) Yijian de Han], issued January 30, 2012, http://www.es.org.cn/c/cn/news/2011-02/15/news_1401.html (has links to draft, the explanation of the draft, and the list of organizations that received this letter soliciting suggestions).
81. National Development and Reform Commission, "Major Fixed Asset Investment Project Social Stability Risk Assessment Provisional Measures," August 16, 2012.
82. "Regulations on the Registration and Management of Social Organizations" [Shehui tuanti dengji guanli tiaoli], issued and effective October 25, 1998, arts. 3, 6, 9. Social organizations refer to non-profit organizations voluntarily created by Chinese citizens, including technical and professional associations. "Temporary Regulations on the Registration and Management of Private, Non-Enterprise Units" [Minban feiqiye danwei dengji guanli zanxing tiaoli], issued September 25, 1998, effective October 25, 1998, arts. 3, 5. Private, nonenterprise units are social groups engaged in nonprofit social services that provide services and utilize state funds, including museums, some hospitals, and schools. "Regulations on the Management of Foundations" [Jijinhui guanli tiaoli], issued March 8, 2004, effective June 1, 2004.

83. According to the 1998 Regulations for Registration and Management of Social Organizations, a group must find a government organization in a similar field as a sponsor. Sponsors supervise the groups. Once a group has a sponsor, they can register with the Ministry of Civil Affairs. For more information on the registration process and the difficulties encountered, see Anna Brettell, "The Politics of Public Participation and the Emergence of Environmental Proto-Movements in China," PhD Dissertation (University of Maryland, 2003), Chapter 6.
84. The definition of nongovernment groups (NGOs) in China can vary considerably, so it is difficult to assess the exact number of environmental NGOs. One article cites figures from the All-China Environment Federation—a government-affiliated NGO, as being about 199 in 2005 and 508 at the end of 2008. The author speculates that the numbers continued to grow at an equivalent rate. Amy Westervelt, "The Year of the (Green) Dragon: China's Burgeoning Environmental Movement," Good, December 30, 2012, http://www.good.is/posts/the-year-of-the-green-dragon-china-s-burgeoning-environmental-movement.
85. For a discussion of the regulatory framework for environmental groups in China, information on the development of these groups, and the debate regarding the influence of these groups, see Anna Brettell, "The Politics of Public Participation and the Emergence of Environmental Proto-Movements in China," PhD Dissertation (University of Maryland, 2003), Chapter 6; Anna Brettell, "Environmental NGOs in the People's Republic of China: Innocents in a Co-opted Environmental Movement?" *The Journal of Pacific Asia*, Vol. 6 (2000):27–56; Guobin Yang, "Environmental NGOs and Institutional Dynamics in China," *The China Quarterly*, Vol. 81 (Spring 2005), 46–66; Michael Busgen, "NGOs and the Search for Chinese Civil Society: Environmental Non-Governmental Organizations in the Nujiang Campaign," The Hague, Institute of Social Studies Working Papers Series No. 422 (February 2006); Turner, Jennifer and Zhi Lu, Chapter in "Building a Green Civil Society in China," The State of the World 2006, Special Focus: China and India. The Worldwatch Institute 2006:152–170; Jennifer Turner, "Small Government, Big and Green Society: Emerging Partnership to Solve China's Environmental Problems." *Harvard Asia Quarterly*, VIII, 2 (2004); Jennifer Turner and Fengshi Wu, eds, *Green NGO and Environmental Journalist Forum: A Meeting of Environmentalists from Mainland China, Taiwan and Hong Kong*. Washington, DC: Woodrow Wilson International Center for Scholars (2004); Fengshi Wu, "Environmental Activism in China: Fifteen Years in Review, 1994–2008," Harvard-Yenching Institute Working Paper Series (2009), http://www.harvard-yenching.org/sites/harvard-yenching.org/files/WU%20Fengshi_Environmental%20Civil%20Society%20in%20China2.pdf; Fengshi Wu, "New Partners or Old Brothers? GONGOS in Transnational Environmental Advocacy in China," *China Environment Series, Woodrow Wilson Center*, Vol. 5 (2005):45–58; Elizabeth Economy, *The River Runs Black: The Environmental Challenge to China's Future*, 2nd Edition (Ithica, NY: Cornell University Press, 2011); Peter Ho, "Greening without Conflict? Environmentalism, NGOs, and Civil Society in China," *Development & Change*, Vol. 32, No. 5 (2001):893–921; Jonathan Schwartz, "Environmental NGOs in China: Roles and Limits," *Pacific*

Affairs, Vol. 77, No. 1 (Spring 2004):28–49; The-Chang Lin, "Environmental NGOs and the Anti-Dam Movements in China: A Social Movement with Chinese Characteristics," *Issues and Studies*, Vol. 42, No. 4 (December 2007):149–184; Yiyi Lu, "Environmental Civil Society and Governance in China." Asia Programme Briefing Paper for Chatham House (August 2005); Caroline Cooper, "'This Is Our Way In': The Civil Society of Environmental NGOs in South-West China," *Government and Opposition*, Vol. 41, No 1. (2006):109–136; Carlos W. Lo and Sai Wing Leung, "Environmental Agency and Public Opinion in Guangzhou: The Limits of a Popular Approach to Environmental Governance," *The China Quarterly*, Vol. 163 (2000), pp. 677–704; Lu Hongyan, "Bamboo Sprouts After the Rain: The History of University Student Environmental Associations in China," *China Environment Series, Woodrow Wilson International Center*, (2004):55–66; Katherine Morton, *International Aid and China's Environment, Taming the Yellow Dragon, Routledge Curzon Studies on China in Transition*, New York: Taylor & Francis, 2006; Tony Saich, "Negotiating the State: The Development of Social Organizations in China," *The China Quarterly* (2000); Jane Sayers, "Environmental Action as Mass Campaign," *China Environment Series*, 5, Woodrow Wilson Center (2003); Judith Shapiro, *Judith, Mao's War against Nature*. Cambridge: Cambridge University Press, 2001.

86. For examples of actions by environmental groups that had some influence on agendas, policies, or projects, see "Congressional-Executive Commission on China, 2009 Annual Report," 199–200, http://www.cecc.gov/pages/annualRpt/annualRpt09/CECCannRpt2009.pdf; "Congressional-Executive Commission on China, 2010 Annual Report," 156, http://www.cecc.gov/pages/annualRpt/annualRpt10/CECCannRpt2010.pdf; "Congressional-Executive Commission on China, 2011 Annual Report," 146–147, http://www.cecc.gov/pages/annualRpt/annualRpt11/AR2011final.pdf; "Congressional-Executive Commission on China, 2012 Annual Report," 117–118, http://www.gpo.gov/fdsys/pkg/CHRG-112shrg76190/pdf/CHRG-112shrg76190.pdf

87. "Former Journalist Draws Pollution Map of China," *Xinhua via China Daily*, June 6, 2013, http://www.chinadaily.com.cn/china/2013-06/06/content_16577013.htm. For more information, see "Institute of Public and Environmental Affairs" [Gongzhong Huanjing Yanjiu Zhongzin], English—http://www.ipe.org.cn/En/index.aspx, and Chinese—http://www.ipe.org.cn/index.aspx.

88. State Administration of Foreign Exchange, "Notice of the State Administration of Foreign Exchange on Issues Concerning the Administration of Foreign Exchange Donated to or by Domestic Institutions" [guojia waihui guanli ju guanyu jingnei jigou juan zeng waihui guanli youguan wenti de tongzhi], December, 25, 2009.

89. State Administration of Foreign Exchange, "Notice of the State Administration of Foreign Exchange on Issues concerning the Administration of Foreign Exchange Donated to or by Domestic Institutions" [guojia waihui guanli ju guanyu jingnei jigou juan zeng waihui guanli youguan wenti de tongzhi], December, 25, 2009, Item 2.

90. Emma Graham-Harrison, "New Finance Rules Add to Squeeze on China NGOs," *Reuters*, March 12, 2010. Some reported no problems accessing

foreign funding, although overall, monitoring of foreign funded groups may have increased.
91. Ministry of Environmental Protection, "Guiding Opinion on Cultivating and Guiding Orderly Development of Environmental Non-Governmental Organizations" [Peiyu yindao huanbao shehui zuzhi youxu fazhan de zhidao yijian], issued December 10, 2010, arts. 2, 10.
92. Ministry of Environmental Protection, "Guiding Opinion on Cultivating and Guiding Orderly Development of Environmental Non-Governmental Organizations" [Peiyu yindao huanbao shehui zuzhi youxu fazhan de zhidao yijian], issued December 10, 2010, art. 10. This may indicate efforts to strengthen Party control over environmental groups.
93. Ministry of Environmental Protection, "Guiding Opinion on Cultivating and Guiding Orderly Development of Environmental Non-Governmental Organizations" [Peiyu yindao huanbao shehui zuzhi youxu fazhan de zhidao yijian], issued December 10, 2010, arts. 2, 10.
94. Ministry of Environmental Protection, "Guiding Opinion on Cultivating and Guiding Orderly Development of Environmental Non-Governmental Organizations" [Peiyu yindao huanbao shehui zuzhi youxu fazhan de zhidao yijian], issued December 10, 2010, art. 10.
95. Raymond Li, "Rights Groups Miss Out on Easing of Registration Rules for NGOs" *South China Morning Post*, March 12, 2013, http://www.google.com/url?sa=t&rct=j&q=&esrc=s&frm=1&source=web&cd=1&cad=rja&ved=0CC4QFjAA&url=http%3A%2F%2Fwww.scmp.com%2Fnews%2Fchina%2Farticle%2F1188742%2Frights-groups-miss-out- easing-registration-rules-ngos&ei=Zl23UZaXLYnI0QHIm4GQBA&usg=AFQjCNFXOkVOzW8nAH66MJUTjVozHmRomQ&sig2=tyhWZhOgvt9NpxWEe243fA&bvm=bv.47534661,d.dmQ.
96. Minzner, Carl. 2009. "Riots and Cover-Ups: Counterproductive Control of Local Agents in China, via SSRN," November 9, http://papers.ssrn.com/sol3/papers.cfm?abstract_id=1502943.
97. Chinese citizens have been submitting petitions to officials as a practice since imperial times. The right to petition, without retribution, is protected in the PRC Constitution (1982). According to Article 41 of the PRC Constitution, "Chinese citizens have the right to criticize and make suggestions to any state organ or functionary. Citizens have the right to make to relevant state organ complaints or charges against, or exposures of, any state organ or functionary for violation of the law or dereliction of duty; but fabrication or distortion of facts for purposes of libel or false incrimination is prohibited. The state organ concerned must deal with complaints, charges or exposures made by citizens in a responsible manner after ascertaining the facts. No one may suppress such complaints, charges and exposures or retaliate against the citizens making them. Citizens who have suffered losses as a result of infringement of their civic rights by any state organ or functionary have the right to compensation in accordance with the law." Sun Weiben, ed., *PRC Administrative Management Encyclopedia* [Zhonghua renmin gongheguo xingzheng guanli dacidian] (Beijing: Renmin Ribao Chubanshe, 1992), 329.
98. For information about early efforts to establish the capacity of environmental protection bureaus to accept and resolve citizens' environmental grievances, see Anna Brettell, "The Politics of Public Participation and the Emergence of

Environmental Proto-Movements in China," PhD Dissertation (University of Maryland, 2003), Chapter 4.
99. State Environmental Protection Administration, "Environment Petition Measure" [Huanjing Xinfang Banfa], issued June 24, 2006, Article 6, http://www.cecc.gov/pages/virtualAcad/index.phpd?showsingle= 186288. This measure is based upon a revised national regulation regarding petitioning that authorities released in 2005.
100. State Environmental Protection Administration, "Environment Petition Measure" [Huanjing Xinfang Banfa], issued April 29, 1996, http://www.law-lib.com/law/law_view.asp?id=13270
101. In Guangzhou as early as 1991, authorities responded to citizen complaints regarding street food stalls by instituting stricter regulations. Authorities also strengthened enforcement efforts in response to citizen complaints regarding noise and air pollution. See Wing-Hung Lo, Carlos and Wing Leung Sai. "Environmental Agency and Public Opinion in Guangzhou: The Limits of a Popular Approach to Environmental Governance," *China Quarterly*, No. 163 (September 2000), 677–704. Other evidence suggests that even if the public is pushing for a cleaner environment, attitudes of supervising governmental officials color environmental authorities' approach to enforcement. Carlos W. H. Lo and Gerald Erick Fryxell, "Effective Regulations with Little Effect? The Antecedents of the Perceptions of Environmental Officials on Enforcement Effectiveness in China;" and Carlos W. H. Lo and Gerald Erick Fryxell, "Governmental and Societal Support for Environmental Enforcement in China: An Empirical Study in Guangzhou," 582. See also Mara Warwick, "Environmental Information Collection and Enforcement at Small-scale Enterprises in Shanghai: The Role of the Bureaucracy, Legislatures, and Citizens," PhD thesis, Stanford University.
102. For descriptions and analysis of environmental disputes through time, see Anna Brettell, "The Politics of Public Participation and the Emergence of Environmental Proto-Movements in China," PhD Dissertation (University of Maryland, 2003), Chapter 3.
103. Shi Jiangtao, "Truth About Pollution Still Shrouded by Secrecy," *South China Morning Post*, January 27, 2012.
104. A variety of sources document the role protests have played in determining project outcomes. Jun Jing, "Environmental Protests in Rural China." In Elizabeth Perry and Mark Selden eds., *Chinese Society: Change, Conflict and Resistance* (New York: Routledge, 2000), 143–160; "China's Media Report on Kunming's Environmental Protests. Translation of 5/28 Southern Weekend Article, East by Southeast," May 30, 2013, http://www.eastbysoutheast.com/?p=466; Samuel Wade, "Anatomy of Two Protests: Kunming vs. Chengdu," *China Digital Times*, May 6, 2013, http://chinadigitaltimes.net/2013/05/anatomy-of-two-protests-kunming-vs-chengdu/; "Congressional-Executive Commission on China, 2009 Annual Report," 200–203, http://www.cecc.gov/pages/annualRpt/annualRpt09/CECCannRpt2009.pdf; "Congressional-Executive Commission on China, 2010 Annual Report," 156–159, http://www.cecc.gov/pages/annualRpt/annualRpt10/CECCannRpt2010.pdf; "Congressional-Executive Commission on China, 2011 Annual Report," 141–143, http://www.cecc.gov/pages/annualRpt/annualRpt11/AR2011 final.pdf; "Congressional-Executive Commission on China, 2012 Annual

Report," 116–117, http://www.gpo.gov/fdsys/pkg/CHRG-112shrg761 90/pdf/CHRG-112shrg76190.pdf.
105. For a general overview of the development of the evaluation systems for government and officials, see John Burns, "Civil Service Reform in China," *OECD Journal in Budgeting*, Vol. 7, No 1, OECD, 2007, http://www.oecd.org/gov/budgeting/44526166.pdf
106. Central Organization Department, "Distributes Provisional Measure on Implementing Local Leadership and Leading Cadres Assessments" [Zhong Zubu Yinfa Shishi Difang Lingdao Banzi He Ganbu Kaohe Shixing Banfa], July 7, 2006, http://politics.people.com.cn/GB/1027/4567774.html; Li Xuan, "Analysis of the '12th Five-Year' Plan Outline, '12th Five-Year' Plan Strengthens the Guiding Role of Environmental Protection Assessments" ["Shier Wu" Guihua Gangyao Jiedu "Shier Wu" Guihua Qianghua Huanbao Kaohe Yindao Zuoyong], April 26, 2011, http://www.zhb.gov.cn/zhxx/hjyw/201104/t20110426_209733.htm
107. Li Xuan, "Analysis of the 12th Five-Year Plan Outline: Acts as a Guide to Strengthen the Environmental Protection Evaluation System" ["Shier Wu" Guihua Gangyao Jiedu "Shier Wu" Guihua Huanbao Kaohe Yindao Zuoyong], *China Environmental News*, April 26, 2011, http://www.zhb.gov.cn/zhxx/hjyw/201104/t20110426_209733.htm.
108. Notice Regarding Initiation of the Enterprise Environmental Protection Evaluation System Pilot Work (1985) [Guanyu Kaizhan Qiye Huanjing Baohu Kaohe Zhidu Shidian Gongzuo de Tongzhi], in *China Environmental and Natural Resources Policy and Law Compendium*, Zhongxin Publishing Company, 1995, 118.
109. Notice Regarding Initiation of the Enterprise Environmental Protection Evaluation System Pilot Work (1985) [Guanyu Kaizhan Qiye Huanjing Baohu Kaohe Zhidu Shidian Gongzuo de Tongzhi], in *China Environmental and Natural Resources Policy and Law Compendium*, Zhongxin Publishing Company, 1995, 118.
110. Li Xuan, "Analysis of the 12th Five-Year Plan Outline: Acts as a Guide to Strengthen the Environmental Protection Evaluation System" ["Shier Wu" Guihua Gangyao Jiedu "Shier Wu" Guihua Huanbao Kaohe Yindao Zuoyong], *China Environmental News*, April 26, 2011. This article mentions the contract target responsibility system as being linked to a five-year program to clean up the Huai River, http://www.zhb.gov.cn/zhxx/hjyw/201104/t20110426_209733.htm.
111. The district measure may or may not be typical of other areas. Jinshanchu 2012 Environmental Protection Target Responsibility Assessment Document Measure, date accessed March 23, 2013, http://huanbaoju.jinshan.gov.cn/FileUpload/%E9%99%84%E4%BB%B6%EF%BC%9A%E9%87%91%E5%B1%B1%E5%8C%BA2012%E5%B9%B4%E7%8E%AF%E5%A2%83%E4%BF%9D%E6%8A%A4%E7%9B%AE%E6%A0%87%E8%B4%A3%E4%BB%BB%E4%B9%A6%E8%80%83%E6%A0%B8%E5%8A%9E%E6%B3%95(%E4%BF%AE%E6%94%B9)(1).doc
112. Li Xuan, "Analysis of the 12th Five-Year Plan Outline: Acts as a Guide to Strengthen the Environmental Protection Evaluation System" ["Shier Wu" Guihua Gangyao Jiedu "Shier Wu" Guihua Huanbao Kaohe Yindao Zuoyong], *China Environmental News*, April 26, 2011. This article mentions the contract target responsibility system as being linked to a five-year

program to clean up the Huai River, http://www.zhb.gov.cn/zhxx/hjyw/201104/t20110426_209733.htm.

113. State Council, "Environmental Protection Committee Decision Regarding Urban Comprehensive Environmental Improvement Fixed Quantity Evaluation System" [Guowuyuan huanjing Baohu Weiyuanhui Guanyu Chengshi Huanjing Zonghe Zhengzhi Dinglian Kaohe de Jueding], in *China Environmental and Natural Resources Policy and Law Compendium*, Zhongxin Publishing Company, 1995, 342. The information can be found in the annual Environmental Protection Yearbook.

114. For example, see the Ministry of Environmental Protection, "'Five-Year' Urban Quantitative Assessment Criteria and Detailed Implementation Rules and Regulations for Comprehensive Improvement of Urban Environmental Protection" ["Shier wu" Chengshi Huanjing Zonghe Zhengzhi Dingliang Kaohe Zhibiao Ji Qi Shishi Xize (Zhengqiu Yijian Gao)], November 7, 2011, http://www.mep.gov.cn/gkml/hbb/bgth/201111/t20111116_220023.htm (Announcement) http://www.mep.gov.cn/gkml/hbb/bgth/201111/W020111116343313075391.pdf (PDF Appendix with details).

115. Ministry of Environmental Protection, "'Five-Year' Urban Quantitative Assessment Criteria and Detailed Implementation Rules and Regulations for Comprehensive Improvement of Urban Environmental Protection" ["Shier wu" Chengshi Huanjing Zonghe Zhengzhi Dingliang Kaohe Zhibiao Ji Qi Shishi Xize (Zhengqiu Yijian Gao)], November 7, 2011, item 2.3, http://www.mep.gov.cn/gkml/hbb/bgth/201111/t20111116_220023.htm (Announcement) http://www.mep.gov.cn/gkml/hbb/bgth/201111/W020111116343313075391.pdf (PDF Appendix with details).

116. Ministry of Environmental Protection, "'Twelfth Five-Year' Urban Quantitative Assessment Criteria and Detailed Implementation Rules and Regulations for Comprehensive Improvement of Urban Environmental Protection" ["Shier wu" Chengshi Huanjing Zonghe Zhengzhi Dingliang Kaohe Zhibiao Ji Qi Shishi Xize (Zhengqiu Yijian Gao)], November 7, 2011, p. 61, http://www.mep.gov.cn/gkml/hbb/bgth/201111/t20111116_220023.htm (Announcement) http://www.mep.gov.cn/gkml/hbb/bgth/201111/W020111116343313075391.pdf (PDF Appendix with details).

117. Ministry of Environmental Protection, "'Eleventh Five-Year' Urban Quantitative Assessment Criteria and Detailed Implementation Rules and Regulations for Comprehensive Improvement of Urban Environmental Protection" ["Shiyi wu" Chengshi Huanjing Zonghe Zhengzhi Dingliang Kaohe Zhibiao Ji Qi Shishi Xize], http://www.zhb.gov.cn/info/gw/huanban/200603/t20060322_75278.htm (Announcement) http://www.mep.gov.cn/image20010518/6341.pdf (PDF of targets)

118. For additional discussion of this system, see Alex Wang, "The Search for Sustainable Legitimacy: Environmental Law and Bureaucracy in China," *Harvard Environmental Law Review*, Vol. 37 (Forthcoming 2013, draft via SSRN, last updated May 21, 2013), http://papers.ssrn.com/sol3/papers.cfm?abstract_id=2128167&download=yes

119. John Burns and Zho Zhiren, "Performance Management in the Government of the People's Republic of China: Accountability and Control

in the Implementation of Public Policy," *OECD Journal on Budgeting*, Vol. 2010/2, OECD, 2010, 2. Burns and Zho provide one of the rare looks into the actual weights given to various criteria.

120. The Chinese names of the measures that regulate these systems were passed by the Central Party Organization Department: 《地方党政领导班子和领导干部综合考核评价办法(试行)》, 《党政工作部门领导班子和领导干部综合考核评价办法(试行)》, and 《党政领导班子和领导干部年度考核办法(试行)》 For information on these measures, see "Questions and Answers Central Organization Department Cadre Assessment Mechanisms Opinions and Three Measures" (full text) [Zhongzhibu Jiu Ganbu Kaohe Jizhi Yijian Ji Sange Banfa Dawen (Quanwen)], *China Net*, October 29, 2009, http://www.china.com.cn/policy/txt/2009-10/29/content_18795210.htm

121. For a more in-depth description of the development and the demise of the Green GDP concept, see Alex Wang, "The Search for Sustainable Legitimacy: Environmental Law and Bureaucracy in China," *Harvard Environmental Law Review*, Vol. 37, (Draft via SSRN, last updated May 21, 2013).

122. State Council, "Decision Regarding Implementing the Concept of Scientific Development in Strengthening Environmental Protection" [Guowuyuan guanyu luoshi kexue fazhanguan jiaqiang huanjing baohu de jueding], issued December 13, 2005, http://www.gov.cn/zwgk/2005-12/13/content_125680.htm

123. Alex Wang provides a detailed description of one of the environmental evaluation systems associated with the five-year plans. For more information, see Alex Wang, "The Search for Sustainable Legitimacy: Environmental Law and Bureaucracy in China," *Harvard Environmental Law Review*, Vol. 37 (Forthcoming 2013, draft via SSRN, last updated May 21, 2013), 40–45. http://papers.ssrn.com/sol3/papers.cfm?abstract_id=2128167&download=yes

124. National People's Congress, Outline of the 10th Five-Year Economic and Social Development Plan [Guojia huanjing baohu "shiwu" guihua de tongzhi], issued March 15, 2001, Part III, Chapter 12, Section 2, http://www.people.com.cn/GB/shizheng/16/20010318/419582.html

125. National People's Congress, "Outline of the 10th Five-Year Economic and Social Development Plan" [Guojia huanjing baohu "shiwu" guihua de tongzhi], issued March 15, 2001, Part III, Chapter 12, Section 2, http://www.people.com.cn/GB/shizheng/16/20010318/419582.html

126. For links to all of the paragraphs in the 11th Five-Year Plan that relate to ecological or environmental protection, see Ministry of Environmental Protection, "Outline of the National Economic and Social Development, 11th Five-Year Plan (environmental protection parts)," April 11, 2006, http://gcs.mep.gov.cn/hjgh/sywgh/dsywgh/.

127. National People's Congress, "Outline of the 11th Five-Year Economic and Social Development Plan" [Guojia huanjing baohu "shiyiwu" guihua de tongzhi], issued March 16, 2006, Part XIV, Chapter 48, Para. 3, http://news.xinhuanet.com/misc/2006-03/16/content_4309517.htm

128. National People's Congress, "Outline of the 11th Five-Year Economic and Social Development Plan" [Guojia huanjing baohu "shiyiwu" guihua de tongzhi], issued March 16, 2006, Part VI, Chapter 24, Sec. 4, http://news.xinhuanet.com/misc/2006-03/16/content_4309517.htm

129. National People's Congress, "Outline of the 11th Five-Year Economic and Social Development Plan" [Guojia huanjing baohu "shiyiwu" guihua de tongzhi], issued March 16, 2006, Part VI, Chapter 22, Sec. 6, http://news.xinhuanet.com/misc/2006-03/16/content_4309517.htm
130. National People's Congress, "Outline of the 12th Five-Year Economic and Social Development Plan," issued March 2011, Part XVI, Chapter 61, Section 3, "[T]he assessment results (of the comprehensive appraisal assessments) are an important basis in the determination of government leadership adjustments and the selection of officials, as well as (giving out) rewards and punishments."
131. National People's Congress, "Outline of the 12th Five-Year Economic and Social Development Plan," issued March 2011, Part XVI, Chapter 61, Section 3, "Increase the momentum of designating and improving an achievement and effectiveness appraisal assessment system and concrete assessment methods that will promote scientific development and speed up transformation of economic development; weaken appraisal assessments of the completion of objectives and tasks regarding the rate of economic growth, and strengthen comprehensive appraisal assessments of the completion of objectives and tasks directed at... conserving energy, protecting the environmental, and providing basic public services and managing society. The assessment results are an important basis in the determination of government leadership adjustments and the selection of officials, as well as (giving out) rewards and punishments."
132. National People's Congress, "Outline of the 12th Five-Year Economic and Social Development Plan" [Guojia huanjing baohu "shierwu" guihua de tongzhi], issued March 2011, Part XI, Chapter 46, Section 3, http://www.npc.gov.cn/wxzl/gongbao/2011-08/16/content_1665636.htm
133. Ministry of Environmental Protection, "Environmental Protection Work Rules" [Huanjin Baohubu Gongzuo Guize], April 23, 2008, art. 25, http://www.mep.gov.cn/gkml/hbb/bwj/200910/t20091022_174593.htm? keywords=环境保护部工作规则
134. "Yun County Environmental Protection Bureau Administrative Accountability System Implementation Rules" [Yunxian Huanbaoju Xinzheng Wenze Shishi Xize], Yun County Environmental Protection Bureau (Lingcang City, Yunnan province), via 360 Doc, September 3, 2008, http://www.360doc.com/content/12/0526/10/6641075_213786141.shtml, Harbin People's Government, Environmental Protection Bureau Implements "Administrative Accountability Rules" Strengthen Work Style Construction Sees Results [Huanbaoju luoshi "xingzheng wenze guiding" jiaqiang zuofeng jianshe jian chengxiao], February 27, 2012, http://www.harbin.gov.cn/info/news/index/detail/269127.htm; Yunnan Xuanwei "City Environmental Protection Bureau Accountability Measure" [Yuannan Xuanwei Shi Huanjing Baohuju Wenze Banfa], Environmental Protection Information Net, posted August 28, 2009, http://in.sunrise-env.com/show.aspx?id=3035&cid=27
135. "Yun County Environmental Protection Bureau Administrative Accountability System Implementation Rules" [Yunxian Huanbaoju Xinzheng Wenze Shishi Xize], Yun County Environmental Protection Bureau (Lingcang City, Yunnan province), via 360 Doc, September 3, 2008,

http://www.360doc.com/content/12/0526/10/6641075_213786141.shtml

136. Office of the Central Committee of the Communist Party and the Office of the State Council, "Provisional Rules Regarding Implementing Party and Government Leader Accountability" [Guanyu Shixin Dang Zheng Lingdao Ganbu Wenze de Shixixng Guiding], Xinhua, issued July 12, 2009, http://news.xinhuanet.com/politics/2009-07/12/content_11696805.htm. Most ministries and governments are likely to have subsequently issued their own versions of these rules.

137. State Environmental Protection Administration and the Ministry of Supervision, "Provisional Rules on Punishments for Violations of Environmental Laws and Discipline" [Huanjing baohu weifa weiji xingwei chufen zhanxing guiding], issued February 20, 2006, art. 2, http://www.mep.gov.cn/info/gw/huanban/200602/t20060222_74265.htm; "Analysis: Provisional Rules on Punishments for Violations of Environmental Laws and Discipline" [Jiedu: huanjing baohu weifa weiji xingwei chufen zhanxing guiding], *China Net*, February 26, 2006. http://www.china.com.cn/chinese/law/1129661.htm

138. State Environmental Protection Administration and the Ministry of Supervision, "Provisional Rules on Punishments for Violations of Environmental Laws and Discipline" [Huanjing baohu weifa weiji xingwei chufen zhanxing guiding], issued February 20, 2006, arts. 4–9, http://www.china.com.cn/chinese/law/1129661.htm; "Circular Regarding Studying and implementing the Provisional Environmental Protection Provisions," http://www.mep.gov.cn/info/gw/huanban/200602/t20060222_74265.htm.

139. State Environmental Protection Administration and the Ministry of Supervision, "Provisional Rules on Punishments for Violations of Environmental Laws and Discipline" [Huanjing baohu weifa weiji xingwei chufen zhanxing guiding], issued February 20, 2006, http://www.mep.gov.cn/info/gw/huanban/200602/t20060222_74265.htm; "Analysis: Provisional Rules on Punishments for Violations of Environmental Laws and Discipline" [Jiedu: huanjing baohu weifa weiji xingwei chufen zhanxing guiding], *China Net*, February 26, 2006. http://www.china.com.cn/chinese/law/1129661.htm.

140. State Council, "National Plan for Environmental Emergency Response" [Guojia Tufa Huanjing shijian Yingji Yu'an], issued, January 1, 2006, http://zfs.mep.gov.cn/fg/xzhg/200611/t20061121_96328.htm

141. Ministry of Environmental Protection, "Interim Measure for Environmental Emergency Response Plan Management" [Tufa Huanjing Shijian Yingji Yu'an Guanli Zhanxing Banfa], issued September 28, 2010, http://www.mep.gov.cn/gkml/hbb/bwj/201010/t20101009_195330.htm

142. Ministry of Environmental Protection, "Sudden Environmental Incident Information Reporting Measure" [Tufa Huanjing Shijian Xixi Baogao Banfa], issued March 24, 2011, Article 15, http://www.mep.gov.cn/gkml/hbb/bl/201104/t20110425_209683.htm

143. "The Environmental Protection Law of the People's Republic of China," art. 38, http://www.china.org.cn/english/environment/34356.htm

144. Supreme People's Procuratorate, "Provisions on the Criteria for Filing Dereliction of Duty and Rights Infringement Criminal Cases," issued

July 26, 2006, Article 19, http://www.gov.cn/ziliao/flfg/2006-07/27/content_346912.htm.
145. "Water Pollution Law of the People's Republic of China" [Zhonghua Renmin Gongheguo Shui Wuran Fangzhifa], Standing Committee of the National People's Congress, issued May 11, 1984, amended May 15, 1996, amended February 28, 2008, Article 4, http://zfs.mep.gov.cn/fl/200802/t20080229_118802.htm.
146. "Circular Economy Promotion Law of the People's Republic of China" [Zhongguo Gonghe Renmin Gongheguo Xunhuan Jingji Cujinfa], NPC Standing Committee, passed August 29, 2008, effective January 1, 2009, Section 6, art. 49 http://zfs.mep.gov.cn/fl/200809/t20080901_128001.htm (Chinese) http://www.amcham-shanghai.org/NR/rdonlyres/4447E575-58FD-4D8E-BB0F-65B920770DF7/7987/CircularEconomyLawEnglish.pdf (English)
147. Ministry of Environmental Protection, "Measures for the Ex-post Supervision of Environmental Administrative Enforcement" [Huanjing Xingzheng zhifa ho ducha banfa], issued, November 5, 2010, Articles 12, http://www.mep.gov.cn/gkml/hbb/bl/201012/t20101217_198805.htm
148. Ministry of Environmental Protection, "Measures for the Ex-post Supervision of Environmental Administrative Enforcement" [Huanjing Xingzheng zhifa ho ducha banfa], issued, November 5, 2010, Article 13, http://www.mep.gov.cn/gkml/hbb/bl/201012/t20101217_198805.htm
149. State Council, "Opinion Regarding Strengthening Key Environmental Protection Work" [Guowuyuan guanyu jiaqiang huanjing baohu zhongdian gongzuo de yijian], October 17, 2011, item 16, http://www.zhb.gov.cn/ztbd/rdzl/hbgzyj/201110/t20111021_218646.htm. The Opinion calls on authorities to strengthen enforcement of environmental laws and supervision, as well as to establish an "enforcement responsibility system."
150. State Council, "National Environmental Protection '12th Five-Year' Plan" [Guojia huanjing baohu "shierwu" guihua de tongzhi], issued December 15, 2011, Part VIII, Section 1, http://www.gov.cn/zwgk/2011-12/20/content_2024895.htm
151. "Measure on Environmental Supervision" [Huanjing Jiancha Banfa], Ministry of Environmental Protection, issued September 1, 2012, Art. 17, http://www.mep.gov.cn/gkml/hbb/bl/201207/t20120731_234120.htm
152. Congressional-Executive Commission on China, *2013 Annual Report*, October 2013, based on the following sources: "Strengthen Environmental Legislation and Improve Supervision and Management Mechanisms" [Qianghua huanjing fazhi wanshan jianguan jizhi], *Legal Daily*, September 3, 2013, http://www.legaldaily.com.cn/News_Center/content/2012-09/03/content_3813016.htm; Li Jingjing, "CCTV Exposé Shows Frustration of Agencies at Keeping Firms in Check," *South China Morning Post*, September 28, 2012, http://www.scmp.com/news/china/article/1048965/cctv-expos%C3%A9-shows-frustration-agencies-keeping-firms-check. "Airpocalypse Now: China's Tipping Point," *ChinaFile*, February 6, 2013. Alex Wang notes that "hard targets are coupled with insufficient monitoring." http://www.chinafile.com/airpocalypse-now-china-tipping-point.) Martin, Paul, Li Zhiping, and Michael Faure

and Hao Zhang, "Toward a More Effective Environmental Criminal Law in China," in Qin Tianbao, Anel Du Plessis, Yves Le Bouthillier, eds. *Environmental Governance and Sustainability*, Northampton, MA: Edward Elgar Publishing, 2012, pp.118–122.
153. "Market-based Policy Instruments for Water Pollution Control in China," Draft Final Report, submitted to the Asian Development Bank and the Ministry of Environmental Protection, May 2010, 47–49, http://www.google.com/url?sa=t&rct=j&q=%20operating%20permits%20polluting%20 enterprises%20china&source=web&cd=8&cad=rja&ved=0CFcQFjAH&url= http%3A%2F%2Fhosted.comm100.com%2Fknowledgebase%2FDownload_ ArticleAttachment.aspx%3Fid%3D100046%26siteid%3D88094&ei= zw25Ud6_JNWw4APp84HgBQ&usg=AFQjCNHfJcfvimYM0I9aGGO7 HUTBY-iiwg&bvm=bv.47810305,d.dmg
154. Michael Faure and Hao Zhang, "Environmental Criminal Law in China: A Critical Analysis," Environmental Law Institute, via US Environmental Protection Agency (2011), http://www.epa.gov/ogc/china/faure.pdf; "Time to Enforce China's Environmental Law," an interview with Wang Canfa, China Water Risk, March 8, 2013, http://chinawaterrisk.org/interviews/time-to-enforce-china-environmental-law/. Both assert that the revisions constitute a step forward but note that other countries apply criminal statutes to a wider range of behaviors.
155. For an in-depth look at public interest cases brought by procuratorates in China, see Mei Hong and Yin Yanjie, "A Feasible Approach to Environmental Public Interest Litigation: the People's Procuratorate as Plaintiff," in Qin Tianbao, Anel Du Plessis, Yves Le Bouthillier, eds. *Environmental Governance and Sustainability* Northampton, MA: Edward Elgar Publishing, 2012, pp. 135–147.
156. Anna Brettell, "Channeling Dissent: The Institutionalization of Environmental Complaint Resolution," in Peter Ho and Richard Louis Edmonds, eds., *China's Embedded Activism*, London, England: Routledge Press, 2008, pp. 137–138.
157. See, e.g., Alex Wang, Sara Imperiale and Wang Pianpian, "Participation, and Environmental Justice: A Comparative Study of China and the United States" Vermont University, 2011–2012, http://www.vermontlaw.edu/Documents/China%20Program/Sara%20Imperiale%20Final%20China%20 JRP.pdf. "Strengthen Environmental Legislation and Improve Supervision and Management Mechanisms" [Qianghua huanjing fazhi wanshan jianguan jizhi], *Legal Daily*, September 3, 2013, http://www.legaldaily.com.cn/News_Center/content/2012-09/03/content_3813016.htm; National People's Congress, "Five Representatives from National People's Congress Organs Answer Journalists' Questions about NPC Work" [Quanguo renda jiguan wuwei fuzeren jiu renda gongzuo da jizhe wen], March 10, 2013 http://www.npc.gov.cn/npc/dbdhhy/12_1/2013-03/10/content_1773886.htm; Ministry of Environmental Protection, "Bulletin Regarding National Environmental Impact Assessment Mechanism Enforcement Review" [Huanjing baohubu tongbao quanguo huanjing yingxiang pingjia jigou zhifa jiancha qingbao], January 28, 2013, http://www.mep.gov.cn/gkml/hbb/qt/201301/t20130128_245608.htm; Chen Yuanyuan, "Need To Raise Speed and Quality for Open Information" [Xinxi gongkai xuyao tisu tizhi], *China*

Environment Net, April 3, 2013, http://www.cenews.com.cn/xwzx/hjyw/201304/t20130402_739268.html; Li Jingjing, "Ex-minister Blames China's Pollution Mess on Lack of Rule of Law," *South China Morning Post*, January 21, 2013, Qu Geping, the former minister of environmental protection blamed the overemphasis on economic growth when he told an international reporter that "...governments have done far from enough to rein in the wild pursuit of economic growth..." He linked unchecked economic development to the "rule of men" and not the "rule of law." "How to Handle the Problem of Grassroots Enforcement?" [Ruhe chuli jizeng zhifa nanti], *China Environmental News*, September 10, 2012, http://www.scmp.com/news/china/article/1132566/ex-minister-blames-chinas-pollution-mess-lack-rule-law.
158. "One-Vote Veto" Used Arbitrarily, "Evaluation System Needs Rectification" ["Yipiao Foujue" Bei Lanyong, Kaohe Zhidu Xu Jiupian], *China News Weekly*, via Xinhua Daily Telegraph, January 30, 2013, http://news.xinhuanet.com/mrdx/2013-01/30/c_132137604.htm.
159. See, e.g., Xu Nan and Zhang Chun, "The Case for Growth in Western China Could Cause 'Huge Surge in Pollution.'" *China Dialogue*, January 23, 2013, http://www.chinadialogue.net/article/show/single/en/5632-The-chase-for-growth-in-western-China-could-cause-huge-surge-in-pollution-; Susan Shifflett, "Paradigm Shift in Chinese Environmental Sector Needed, Says Activist Wang Canfa," Environmental Change and Security Program (NewSecurityBeat Blog), March 21, 2013, http://ht.ly/jhton. (CECC Annual Report, 2012, 114).
160. "Toward an Environmentally Sustainable Future: Country Environmental Analysis of the People's Republic of China," *Asian Development Bank*, August 2012, 95, http://www.adb.org/sites/default/files/pub/2012/toward-environmentally-sustainable-future-prc.pdf. The State Environmental Protection Administration reportedly said that the new Ministry of Environmental Protection needed 700–800 staff members (instead of its then 368).

CHAPTER 3

DOES CADRE TURNOVER HELP OR
HINDER CHINA'S GREEN RISE?
EVIDENCE FROM SHANXI PROVINCE

Sarah Eaton and Genia Kostka

1. INTRODUCTION

In the last decade Beijing has presided over an ambitious effort to lay the foundations of a green economy. The clear message sent by China's top leadership is that environmental constraints and resource scarcity imperil future economic growth and social harmony in China. Greening growth initiatives—here defined as measures promoting the use of less resources for economic growth—aim to help China change lanes from a heavily polluting, growth-at-any-cost model to a resource-efficient and low-carbon model. Yet, Beijing's various efforts to steer China in the direction of greener growth have often met with resistance during policy implementation, when sub-national leaders take center stage. This study argues that, among a range of factors that shape the local politics of greening growth, the frequent turnover of local leaders contributes to the green implementation gap.

This research addresses a growing literature exploring why the central government's environmental aspirations are not always mirrored at local levels. Previous contributions have illuminated the significance of bureaucratic fragmentation, decentralization, limited institutional capacity and incentives embedded in the cadre evaluation system in deflecting and diluting Beijing's green growth initiatives. To this list we add the high cadre turnover. We argue that the typically brief tenure of local leaders exists in some tension with the weighty leadership demands of greening growth, a core priority of China's two most recent "green" five-year plans. The state-led greening growth envisioned in China's 11th and 12th Five Year Plans (FYPs) requires far-sighted and locally rooted leaders. Since local leaders tend to cycle in and out of the leadership group (*lingdao banzi* 领导班子) at three- or four-year intervals, they are often ill-equipped and insufficiently incentivized to steer their localities in the direction set by Beijing.

The findings are based on extensive fieldwork in Shanxi province during 2010 and 2011. We selected two coal-dependent localities within Shanxi currently in the midst of green growth transformations with contrasting experiences of cadre rotation in the local leadership group: Datong city (*Datong shi* 大同市) and Xiaoyi city (*Xiaoyi shi* 孝义市). For most of the last two decades, Datong has had a high rate of turnover in the local leadership group. Xiaoyi, by contrast, is characterized by an unusually continuous and stable leadership. In total, the case study analysis draws on 45 interviews in Shanxi. The majority of interviews were conducted with leading officials in the Chinese Communist Party Organization Department, Environmental Protection Bureau (EPB), Development and Reform Commission (DRC), and Economic Commission as well as with industrial enterprise managers. Interviews were semistructured and provided an understanding of overall greening growth initiatives in Datong and Xiaoyi and shed light on the significance of the two localities' distinct experiences of cadre turnover.

The argument starts from an overview of China's national green growth ambitions before turning to a brief review of the literature on the green implementation gap in China. The next section explores the rationale of China's post-1978 cadre rotation system and summarizes its perceived advantages and disadvantages for policy implementation. In the subsequent section we narrate the case studies and unpack the complex factors giving shape to green growth initiatives in Datong and Xiaoyi. The analysis focuses particularly on the role of cadre turnover in shaping implementation outcomes. The final section discusses the lessons and policy implications suggested by this research.

2. Greening Growth

Over the last decade, China has begun to form a national answer to the challenges of domestic resource scarcities and environmental degradation. The 11th and 12th Five Year Plans reflect the national policymaking priority of shifting toward a more sustainable and resource-efficient growth model. The 11th FYP established conservation, efficient use of resources and economic transformation in the interests of sustainable development as a "basic national policy" (*jiben guoce* 基本国策) (Casey and Koleski 2011). The 12th FYP has deepened Beijing's vision for upgrading and restructuring the economy by offering specific guidance on how to shift to higher value-added manufacturing, improve the conservation of energy and resources, and develop service industries. It also identifies seven strategic emerging industries (SEIs), state agencies that are to nurture in order to aid the shift to higher-value industries and sustainable growth.[1] The core priority of the recent "green" plans is economic transformation based on industrial upgrading in the second sector and expansion of the third sector. Table 3.1 summarizes the main "hard," literally restricted (*yueshuxing* 约束性), and "soft," expected (*yuqixing* 预期性) targets of the 11th and 12th FYPs. In addition to the national plans, a mix of laws, taxes, industrial policies, guidelines, and regulations are also propelling the switch to a resource-efficient and low-carbon growth path.[2]

2.1. The Green Implementation Gap

Yet Beijing's expansive vision does not always find a receptive audience at lower levels. Indeed, recently reported figures show that many of Beijing's green growth initiatives are implemented only selectively across China. For example, most targets related to transforming China's economic development pattern were only partially met during the 11th FYP period (2006–2010). Officially, the only hard target not fulfilled was the goal of reducing energy intensity per unit of GDP by 20 percent as compared to 2005 levels (the actual recorded reduction was 19.1 percent). But, anecdotally, among the localities reported to have met the energy intensity target, there may well have been falsification of these figures.[3] Among the softer targets related to economic restructuring in the 11th FYP, fully three were unmet: (1) increasing service sector as a percentage of GDP, (2) service sector as a percentage of employment, and (3) R&D as a percentage of GDP (see Table 3.1).

Moreover, plan fulfillment is not always a reliable indicator of true progress on greening growth. Many of the environmental targets in the 11th FYP were implemented late in the game and were not achieved by making lasting changes. For example, in some localities, mandatory energy intensity targets were fulfilled only at the last minute using extreme and sometimes socially harmful measures. Such energy-saving measures included cutting electricity to hospitals, homes, and rural villages. During the last quarter of the 11th FYP, local governments also used "sleeping management" (*xiumian guanli* 休眠管理) to temporarily shut energy-intensive companies (Kostka and Hobbs 2012; Harrison and Kostka 2014). Such drastic 11th-hour practices illustrate that some local leaders choose the path of least resistance in selecting short-term, low-quality approaches to satisfying environmental targets and, while nominally following green directives, are actually putting off the difficult business of changing lanes. Achieving the economic restructuring envisioned in China's green plans will often require a high degree of local policy coordination, significant government expenditure and overcoming opposition from polluting industries and other entrenched interest groups. Previous work sheds light on many of the factors that make the launching of such complex, costly, and long-term initiatives a difficult proposition in many localities.

Table 3.1 Major economic and environmental targets in the 11th FYP and 12th FYP

Targets	11th FYP (2010 target)		2010 Actual	12th FYP (2015 target)	
Economic targets					
Average GDP growth (percent)	7.5	E	11.2	7	E
Average GDP growth per person	6.5	E	10.6	N/A	
Service sector as percent of GDP	43.3	E	43 (*not met*)	47	E
Service sector as percent of total employment (percent)	35.3	E	34.8 (*not met*)	N/A	
Urbanization (percent)	47	E	47.5	51.5	E
R&D as percent of GDP	2	E	1.75 (*not met*)	2.2	E
Patents per 10,000 people	N/A		1.7	3.3	E
Strategic industry as a percent of GDP	N/A		N/A	+8.0	E
Environmental targets					
Reduction in energy intensity per unit of GDP (percent)	20	R	−19.1 (*not met*)	−16	R
Reduction in carbon intensity per unit of GDP (percent)	N/A		N/A	−17	R
Non-fossil fuel in primary energy mix (percent)	N/A		N/A	11.4	R
Major pollutants (percent)	COD: −10 SO$_2$: −10; Ammonia: N/A Nit. oxide: N/A	RR	COD: −12.45 CO$_2$: −14.29	COD: −8 SO$_2$: −8 Ammonia: −10 Nit. oxide: −10	RRRR
Forest coverage (percent)	20	R	20.36	21.66 (or 14.3 trillion cubic meters)	R
Reduction of water consumption per unit of value-added of industrial output (percent)	−30	R	−36.7	−30	R
Increase of water efficiency coefficient in agricultural irrigation	0.5	E	0.5	0.53	E
Farmland reserves (million hectares)	120	R	121.2	121.2 (or 1.8 billion mu)	R
Comprehensive utilization rate of industrial solid waste (percent)	60		69	N/A	

Note: Population targets are excluded. R refers to a restricted, hard target; E refers to an expected, soft target; N/A indicates that no target was stated in the respective FYP.
Source: Adapted from Appendix 2 from Casey and Koleski (2011, 15–16).

2.2. Determinants of the Green Implementation Gap

The existing literature has uncovered a variety of factors that contribute to the green "implementation gap." First, while national ministries set the overall direction and long-term goals, these aspirations are often diluted as they pass through the institutions that make up China's fragmented vertical and horizontal (*tiao tiao kuai kuai* 条条块块) governance structure (Lieberthal and Oksenberg 1988). A second view explains the gap between national policies and local practices with reference to China's decentralized governing structure, which allows local officials to be choosy about which national policies to faithfully implement and which to put on the back burner (Economy 2004). Other scholars reject the decentralization premise and argue that the center still wields substantial power; this perspective holds that formal constraints imposed by the central government should be held to account for shortcomings in environmental policy implementation (Ran 2013). And yet another branch of literature points to the frequent divergence of national and local interests, as local leaders tend to place greater emphasis on economic and industrial development than on environmental and resource concerns (Van Rooij 2006; Kostka and Hobbs 2012, 2013). For example, in defiance of the center's 7 percent GDP growth target in the 12th FYP, provincial FYPs for the period show that only five of the 31 provinces have set growth rate targets below 10 percent, indicating that provinces continue to place most emphasis on growth-focused development (Huang 2011).

Another branch of research sees selective policy implementation as largely a function of incentives embedded in the cadre responsibility and evaluation systems (O'Brien and Li 1999). O'Brien and Li (1999) argue that "street-level bureaucrats" are incentivized to prioritize projects that they think will advance their career. In angling for promotion, leading cadres appeal not to central bureaucrats but, first and foremost, to leaders one administrative level above who bear responsibility for personnel decisions. Local officials may select showy "political accomplishment projects" (*zhengji gongcheng* 政绩工程) such as extravagant construction projects in the effort to set themselves apart from their competitors in other localities (Cai 2004). For careerist officials, choosing *zhengji gongcheng* is seen as a tried and true tactic for currying favor with upper-level authorities (Guo 2009:623).

In the process of selective implementation, carrying out less measurable environmental policies may well fall to the bottom of a leading cadre's implicit list of priorities. To be sure, Beijing continually tinkers with the cadre responsibility and evaluation systems in order to incentivize leaders in local governments and state-owned enterprises (SOEs) to carry out the green growth vision outlined in the national planning documents. Yet, the properties of certain environmental policies make them leading candidates for implementation shirking.[4] In China, as in many political systems, there appears to be a structural bias favoring implementation of readily measurable goals (O'Brien and Li 1999; Eaton and Kostka 2014). Most targets handed down to the localities—e.g., government revenue targets and caps on the number of local petitioners—orient local leaders toward attaining readily measurable goals. By contrast, some aspects of the nascent green growth agenda are far more elusive. Whereas assessment of a locality's performance in raising forest coverage rates (one of the notable successes of the 11th FYP) is relatively straightforward, technological and organizational challenges make measurement of energy and carbon intensity extremely difficult. In combination with weak checks in the environmental reporting system, it is not hard to see why, under these circumstances, local leaders might be tempted to simply falsify reports on less readily measurable elements of the green growth agenda.

To these factors contributing to the green implementation gap identified in previous work, we add the high rate of turnover among local leaders. Previous research suggests that incentives in the cadre evaluation system interact with the typically short time horizons of local Party secretaries and mayors to encourage local leaders to select the path of least resistance in local implementation of greening growth initiatives (Eaton and Kostka 2014). Given a strong interest in selecting projects that will place realized political accomplishments on their CVs by the end of their term, policies that are more risky or need a longer time to complete are sometimes sidelined. As such, cadres' incentives to secure short-term goals are sometimes starkly at odds with the center's green growth mandate, many aspects of which are not realizable in the short term. High cadre turnover can also frustrate green growth implementation in more prosaic ways since frequently rotated officials spend much of their time simply getting up to speed in their new localities (Eaton and Kostka 2014).

3. Pros and Cons of High Cadre Turnover

Whereas the significance of election cycles is well established in the study of democratic politics, the policy impact of local leadership cycles in China—which actually move at a faster rate than their democratic counterparts—remains curiously underexplored. In a previous study we outlined the basic characteristics and policy background of cadre turnover and discussed intended and unintended consequences of this system for local implementation of greening growth policies in general (Eaton and Kostka 2014). In this earlier study, we presented primary and secondary data that the two pillars of local leading groups (*lingdao banzi* 领导班子), Party secretaries and mayors, are typically rotated to a new locale every three to four years. This cadre turnover takes the form of promotion, lateral rotation, and, much less frequently, demotion.[5] The center views institutionalized cadre switching as a means of exercising control over the localities and limiting local leaders' opportunities to engage in corruption and policy defiance. Analyzing the effects of cadre turnover at local levels, we previously showed that while there are some benefits associated with the system, there are also some notable downsides for environmental policy implementation.

On the one hand, cadre turnover is seen to have a number of benefits for policy implementation. First, short leadership cycles can, indeed, aid policy implementation through reducing coordination problems by bridging horizontal and vertical administrative gaps. Since 2005, a number of central and provincial-level Party and government organization have rolled out various rotation schemes with the explicit aim of improving coordination and cooperation. For example, in 2009, the State-owned Assets Supervision Administration Commission (SASAC) set up a rotation plan with the central SOEs under its watch, which envisioned rotating officials as crucial bridges between the supervisor and the supervised and, therein, as helpful to the cause of SOE reform. Second, cadre turnover can serve to disseminate effective governing practices from place to place. For example, Qiu He brought his heavy-handed approach to stamping out crime and government corruption, honed in leadership posts in Suqian City (Jiangsu) to his current position as Party secretary of Kunming city in Yunnan province.

Yet, although the cadre rotation system has many advantages for environmental policy implementation, there are also acknowledged downsides to the system in its current form. In particular, the expectation of a short stay may encourage leaders to place most emphasis on policies that will bear fruit quickly and to tacitly sideline policies with a longer time to maturity. Frequent post changes may also limit local leaders' capacity to comply with central directives given friction costs associated with adapting to new localities. Moreover, institutionalized post-shuffling of local leading cadres can result in damaging discontinuity in local development initiatives.

The short tenure of leaders combined with their interest in accumulating political accomplishments (*zhengji*) can encourage short-sighted behavior and inhibit the formation of long-term development plans. Figure 3.1 suggests how the personal incentives of promotion-hopeful cadres might affect their prioritization of different green growth policies. While there are likely to be many factors involved in the selective implementation of greening growth, we see the *zhengji* value along with the time frame of a given initiative as particularly important in steering cadres toward some policies and away from others.[6] In the context of the recent green plans, hard targets (e.g., forest coverage rate) would tend to be assigned a higher *zhengji* value than soft targets (e.g., increasing service sector as a proportion of GDP). The "project time to maturity" axis implies that projects that deliver tangible results within the leaders' tenure (i.e., within 3–4 years) will tend to be selected over projects with a longer time frame. The matrix shows how these *zhengji* and time considerations interact in the process of selective policy implementation. It suggests that, in allocating their energy and scarce resources to competing projects and fulfilling environmental policy mandates, leading officials are likely to pick green projects with both high political accomplishment (*zhengji*) value and a high probability of producing results within their own tenure cycle (quadrant 3). By contrast, projects that are not seen to enhance a cadre's chances of promotion and that take a long time

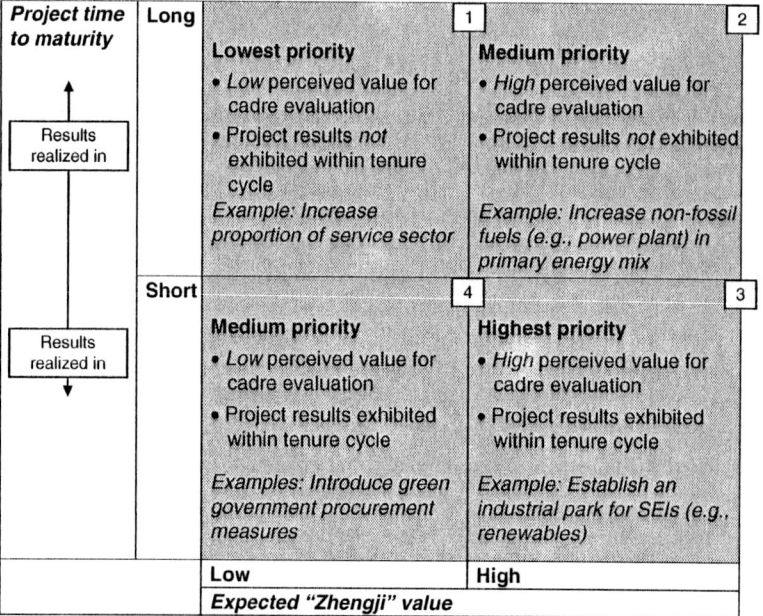

Figure 3.1 Project prioritization matrix: The view of a promotion-seeking official
Source: Authors.

to produce results will tend to be ranked at the bottom of the prioritization list (quadrant 1). Projects that make cadres stand out in their peer group (high *zhengji*) but without deliverables at the end of the tenure period might also be sidelined, depending on how much credit he or she can expect to gain from the long-term initiative (quadrant 2). Finally, projects with low perceived value for cadre evaluations but with realizable results in the short term might be picked to highlight the wide range of activities undertaken in a particular cycle (quadrant 4).

While previous research outlines the pros and cons of high cadre turnover on local governance, there is a lack of detailed case study analysis on the effects of the cadre rotation system. As our subsequent case studies show, when facing costly and complex policy mandates, short-staying leaders may choose the path of least resistance in selecting quick, low-quality approaches to the implementation of environmental policies while more locally rooted leaders with strong links to local industry may use the benefits of embeddedness to ease implementation.

4. Case Studies: Datong and Xiaoyi—Variation in Cadre Turnover

Our case studies focus on two localities in Shanxi province currently in the midst of greening growth. Datong is a municipal-level city (*di ji shi* 地级市)

while Xiaoyi city is a county-level city (*xian ji shi* 县级市) under the administration of Lüliang municipality. While the implementation contexts in counties and municipalities differ in important ways, our primary aim is not to directly compare the two but, given the exploratory nature of this research, to assess the significance of cadre turnover within each case study. Despite their different administrative designations, the two localities' economies and coal extraction volume are comparable in size. Reported GDP in 2009 for Datong and Xiaoyi were 33 billion RMB and 20 billion RMB, respectively. Datong and Xiaoyi are similar in that they are well known for coal production and lack a diversified industrial structure and are home to a large proportion of energy-intensive enterprises. In 2007, Datong produced 24.6 million tons of coal and Xiaoyi 25.1 million tons. The cities differ in terms of population and average income distribution. Datong has a population of 3.2 million and Xiaoyi just 0.4 million. GDP per capita was 19,774 RMB in Datong and 45,538 RMB in Xiaoyi (the highest in Shanxi) in 2009 (Shanxi 2010 Statistical Yearbook, Tables 18.13, 19.1, 19.3).

The two localities have had markedly different experiences of cadre turnover. Leadership in Datong has been fairly unstable, with Party secretaries staying an average 3.73 years and mayors an average 2.45 years with the accumulated average time served in either position summing to 3.93 years. Prior to the 2008 arrival of a charismatic mayor who pledged, at a minimum, to serve out his five years in full, the turnover of Party secretaries and mayors every three to four years made Datong a fairly typical city in China. By contrast, in Xiaoyi, local leaders tend to stay in office for much longer. Recent Party secretaries have served an average of 8.29 years and mayors an average of 6.22 years (see Table 3.2). It is also common practice in Xiaoyi that mayors stay on to become Party secretaries, and, consequently, the accumulated average serving time in both positions is 9.31 years. The large discrepancy between leadership patterns in Datong and Xiaoyi presented a reason to investigate the possible impact of high/low cadre turnover on local leaders' ability and motivation to green local economies.

5. Datong: Geng Yanbo lai le!

We say that Datong has only had two great mayors. One is Tuo Bahong (founder of the Northern Bei dynasty who chose Datong as the capital) and the other Geng Yanbo.[7]

Shanxi's second largest city, Datong, is in the throes of an ambitious and costly effort to shed its title as China's "coal capital" and reinvent itself as a tourist destination. Historical records suggest that as early as the Tang dynasty local residents began mining the 1,800 km² Datong Coalfield for heating and iron smelting purposes and, ever since, the city's fortunes have been tied to the vast deposits of coal underfoot (Fu et al. 2010:1). Datong's two pillar enterprises attest to this legacy. Datong Coal Group, China's second largest coal enterprise, produces about 100 million tons of coal annually and the

Table 3.2 Party secretaries and mayors in Datong and Xiaoyi

		Datong city tenure years			Xiaoyi city tenure years	
Party secretary	Feng Lixiang 丰立祥	02/08–		Zhang Xuguang 张旭光	12/09–	
	Guo Liangxiao 郭良孝	02/06–02/08	2.00	Li Liangsen 李良森	05/01/–12/09	8.58
	Lai Yulong 来玉龙	10/01–02/06	4.33	Zhang Huosheng 张霍生	05/93–05/01	8.00
	Jin Shanzhong 靳善忠	02/00–10/01	1.67			
	Ji Youwei 纪友伟	03/93–02/00	6.92			
Average (1993–)			3.73			8.29
Mayor	Geng Yanbo 耿彦波	02/08–		Guo Jiping 郭继平	12/09–	
	Feng Lixiang 丰立祥	03/06–02/08	1.92	Zhang Xuguang 张旭光	12/01–12/09	8.00
	Guo Liangxiao 郭良孝	06/03–03/06	2.75	Li Liangsen 李良森	06/98–12/01	2.50
	Sun Fuzhi 孙辅智	02/00–06/03	3.33	Cao Jinchu 草金初	04/90–06/98	8.17
	Jin Shanzhong 靳善忠	03/98–02/00	1.92			
	Du Yulin 杜玉林	02/95–03/98	3.08			
	Cheng Buyun 程·云	11/92–02/95	2.25			
	Li Youmei 李有美	09/90–11/92	2.17			
Average (1990–)			2.45			6.22

Sources: Datong and Xiaoyi gazetteers (various years).

coal-fired Datong No. 2 Power Station is a major electricity supplier to the Beijing–Tianjin–Tangshan region. The air and water pollution caused by the predominance of coal in Datong's industrial structure has earned the city the epithet "black hat" (*hei maozi* 黑帽子) for environmental degradation. Indeed, a national 2005 EPB ranking of the environmental quality of 113 Chinese cities found Datong third from the bottom (Ma and Zhang 2006). Besides severe air and water pollution, heavy mining activity has badly deformed much of the local geological environment causing soil avalanches, landslides, and surface cracking, among other problems (Fu et al. 2010:1–4). The heavy environmental toll of Datong's coal-reliance, in addition to dwindling supplies of marketable coal, have long been recognized as a near crisis. However, a succession of lackluster leaders failed to develop any strategy for greening the city's growth.

5.1. Leadership Cycles

The arrival of a new mayor in 2008, Geng Yanbo, marked a decisive break with the past and this charismatic leader has worked to develop new sources of growth in the second and third sectors, though in a risky and highly controversial fashion. Prior to Geng's arrival, Datong had a run of unremarkable leaders with weak incentives to correct the city's coal dependence. Looking at the average tenure of local leaders over the past two decades we find that Datong was fairly typical of Chinese municipalities in having mayors and Party secretaries come and go quickly. One well-placed city official said that the leaders before Mayor Geng had used their time in Datong primarily as an opportunity for advancement:

> Usually, previous leaders stayed for a couple of years and then found a chance to get promoted. They used Datong as a springboard. The Party secretary of Datong is a position from which it's easy to get promoted because there are many coal bosses here. That means they can get a lot of bribes and use them to bribe the upper level government to get promoted. They didn't even need any political achievements ["*zhengji*"] to prove themselves... None of them made any difference. When they left, Datong was the same as when they came. For them, all was good so long as no major problems arose.[8]

The career paths of two recent leaders illustrate this pattern. Guo Liangxiao was Datong's mayor for 2.8 years (2003–2006), then Party secretary for a 2-year stint (2006–2008) before being promoted to vice-chairmanship of Shanxi's Chinese People's Political Consultative Conference (CPPCC). Jin Shanzhong followed a similar route. He was Datong mayor for 1.9 years (1998–2000) and then Party secretary for 1.7 years (2000–2001) before moving on to leadership positions in the provincial Shanxi Enterprise Work Committee, SASAC, and the NPC Standing Committee.

Mayor Geng Yanbo cannot be accused of indolence or risk-averse opportunism. Geng's uncompromising leadership style resembles that of China's other prominent "officials with personality" (*gexing guanyuan* 个性官员) such as Party secretary of Chongqing, Bo Xilai and Qiu He, current Party secretary of Kunming (Fewsmith 2006, 2010). Reportedly, from his very first days in office in 2008, Mayor Geng assumed the role of first in command (*yi ba shou* 一把手) much to the displeasure of Datong's current Party secretary, Feng Lixiang. Feng is said to have complained to the provincial authorities about Geng's flouting of rank but the upper levels ultimately backed Geng and, by all accounts, the mayor is now the final authority on all important decisions in Datong.[9] Geng let his colleagues know early on that he would chart a different course:

> When Geng first arrived in Datong, he stopped all the real estate projects that did not meet construction standards. Real estate developers interpreted this as a sign that Geng wanted bribes but they were wrong. Geng directed his subordinates to take the money they offered him and deposit it in a special government fund used for city construction.[10]

While many within Datong government seem to appreciate the vigor Geng brings to his work, his subordinates are also fearful of his bad temper:

> Sometimes Geng swears at or even beats people. At first, officials didn't know him well and came late to meetings. Geng would then tell them "you don't need to come here anymore" and remove them on the spot. If they failed to complete their tasks, they were also removed. Some officials handed in their resignation because there was too much pressure. Geng would refuse them. They had no other choice but to do it well.[11]

Geng's dictatorial style has certainly earned him enemies within Datong government but many ordinary people admire his enthusiasm—he often says he "wants the city to get excited"—as well as his work ethics and his reputation for non-corrupt rule. Geng actively uses the local media to promote this hardworking image and he has even inspired a pop song tribute called "The Soul of Datong" circulating on the internet. Presumably most of those displaced by the grand-scale demolition of old buildings as part of Geng's city revitalization initiatives would be slightly less upbeat about the new mayor and there have been significant numbers of petitioners and "nailheads" (*dingzihu* 钉子户) since Geng's arrival (Honesty Outlook 2011).[12]

A history enthusiast, Geng had a vision for Datong's green growth based on refurbishing and sometimes outright rebuilding Datong's cultural riches in order to attract tourism. In settling on this approach, Geng drew from his experience in previous posts elsewhere in Shanxi. In Lingshi county in Jinzhong municipality, where Geng was vice-secretary of the Party and then mayor (1993–2000), a major project of his was restoration of the Wang Family Mansion.[13] In Yuci (2000–2006), another district under Jinzhong municipality, Geng presided over a large number of controversial construction projects. These included demolition of old buildings and their replacement with new ones as well as refurbishment of Chang Manor and

Yuci Old Town in order to promote tourism.[14] Such was the degree of opposition to these construction projects in Yuci that, at the end of his term, someone left Geng three memorial wreaths. Yet, interviews in both localities conducted in 2010 suggest that, for some, Geng's tenure is now seen with a measure of nostalgia: "People all miss him in Lingshi and Yuci because they benefited from the construction. Especially in Yuci, he is still a hero."[15] Geng also learned in his previous posts that top-to-bottom city remodeling cannot be compressed into the typical term of a municipal mayor or Party secretary. Indeed, when he was Vice-Mayor of Taiyuan (2006–2008) and got word that he would soon be re-posted to Datong, Geng demanded to stay for at least five years knowing that he would need at least this much time to "make a long-term impact" and cement his status as "a leader with vision."[16]

5.2. Greening Growth

The official "three famous, one strong" (*san ming yi qiang* 三名一强) moniker for Datong's development under Geng's leadership envisions Datong as a famous cultural, tourist, and ecological city with a strong economy. "Three famous" indicates the high priority placed on tourism in Geng's greening growth strategy. In his first Government Work Report, Geng declared: "Culture is always a city's means of survival and a source of competitiveness in development. Datong's rich historical relics are non-renewable, irreplaceable and the city's most valuable resources" (Datong Government Work Report 2011). "One strong" refers, in part, to the promotion of non-coal industries in the second sector. Within the city limits, Geng's major cultural projects include renovation of significant Buddhist sites including Huayan and Shanhua temples and reconstruction of the ancient city wall. Geng's cultural restoration initiative extends to select counties under the "body and two wings" approach (*yi ti liang yi* 一体两翼) in which Datong proper is the body and the wings are formed by the Yungang Grottoes (in Datong's Western district) and Mount Heng (in Hunyuan county), one of the five sacred Daoist mountains and home of the Xuankong "Hanging" temple. For Hunyuan county, inclusion in Geng's plans has meant an influx of financial support from the municipality and the central government for tourism-related projects.[17] Investment in restoration of the Yungang Grottoes amounts to 1.7 billion RMB, a combination of funding from the National Heritage Board, the province and the municipality.[18] In addition to large-scale cultural construction projects, Geng's three years in office have also seen huge infrastructure investment in Datong. The total investment in road and bridge construction is said to amount to 6 billion RMB and, in 2010 alone, 38 new roads were completed, bringing the total built in Geng's tenure to 69.[19]

Since Geng's arrival, Datong government has also taken a strong approach to implementation of national policy on coal industry consolidation. Central documents urging the promotion of large-scale coal enterprises have been issued largely because larger mines tend to be safer (due to higher mechanization rates), less polluting, and more efficient in energy use and coal extraction

(IEA 2009:44). In 2008, Geng pledged support for Datong Coal Group's "going big and going strong" (*zuo da zuo qiang* 做大做强) and promised the gradual closure of coal mines with annual production of less than 30 million tons of coal (Datong Daily 2008:10).[20] In the following year, the municipal government led a major industry restructuring, which saw 127 coal mines consolidated into 65. In the process, Datong Coal Group took over eight new mines. The 2009 consolidation had benefits in terms of helping surviving enterprises achieve scale economies and the associated energy savings contributed to Datong's reported achievement of its 11th FYP energy efficiency targets. The bold approach to scaling up Datong's coal industry has, though, come at a great cost in terms of lost revenue. As of June 2011, of the 65 consolidated coal mines, just eight were producing normally, while 17 others are said to be improving their capacity and facilities.[21] The slow-down in coal mining activity since the 2009 consolidation is part of the reason Datong's GDP rank among Shanxi cities slipped to number eight in 2010 from number two in 2005.[22]

Geng has also presided over the development of non-coal industries on the city's outskirts. The city's "one axis, two cities" (*yi shou shuang cheng* 一轴双城) approach refers, first, to the makeover of the old city within Datong and, second, to the establishment of the Yudong New Area to the east of the city. Set up in 2009, Yudong is a new industrial and residential area where the city has pinned its hopes of developing new industries. The area features a 20 km² automotive-focused manufacturing park, two pharmaceuticals parks, a new materials park, and a new energy (solar and wind) park. While interviewees expressed a cautious optimism about the prospects for these new industries to reduce Datong's coal dependence, like many of Geng's initiatives, Yudong New Area has also stirred controversy. Datong municipality directly manages the Yudong New Area but the manufacturing and pharmaceuticals parks are physically located within the boundaries of the poor, revenue-strapped Datong county.[23] Datong municipality reportedly told Datong county officials that in order to lay claim to 20–30 percent of future taxes in these industrial parks, the county would first have to invest 100 million RMB in their construction.[24] To this end, Datong county, which is on an "eating budget" (*chi fan caizheng* 吃饭财政), has already invested 72 million RMB in the new projects.[25] County officials also worry about the pollution these industrial parks will bring to their predominantly agricultural region and are particularly concerned about the impact of waste water from pharmaceuticals factories. In the context of Datong municipality's own growing revenue woes, Yudong is seen by some as an effort to push pollution out of the city while bringing revenue in.[26]

While many see Geng as laying a firm foundation for Datong's industrial transformation, it remains to be seen whether this greening growth strategy is in fact viable. To be sure, there have been notable improvements in the local environment in recent years. In 2010, the EPB's air quality monitoring system showed Datong as having air quality of Level Two or above on 341 days, up from 307 in 2008 (Wang 2010). Cleaner air reflects reduced coal mining activity as well as the municipal government's 500 million RMB investment in district heating and cogeneration facilities to replace highly polluting coal-fired boilers.[27] And Geng's cultural projects may have begun to lure visitors to Datong as tourist numbers have apparently been increasing at a rate of about 20 percent annually in recent years.[28] Reinventing Datong as a tourist destination will likely be an uphill task, however, as many Chinese and foreigners alike still associate the city more strongly with the "black hat" than with Buddhist treasures. Likewise, switching to greener sources of growth will be a slow and difficult process. As a recent Datong DRC report put it, "The situation of relying mainly on coal, electricity and other energy industries for industrial economic growth has not changed much. Improvement of the eco-environment and energy savings and emissions reduction work is still arduous"[29]

Aside from the inherent risks of this tourism-based strategy, Mayor Geng's one-man show has also placed enormous pressure on municipal finances. Under the pressure of a *de facto* five-year term limit, Geng launched many initiatives almost immediately upon taking office and project financing has been pieced together on the fly. Geng's strategy of attracting central SOE investment (*yangqi jintong* 央企进同) has had some success but, on the whole, enterprise investment has played only a minor role in Datong's transformation so far.[30] Typical of his cannot-wait attitude, Geng's reply to recent questions about city construction projects plagued by legal violations was "I do not have time to wait, so demolition cannot wait" (The Beijing News 2011). Some funding for cultural restoration projects has come from the central and provincial governments (e.g., for the Yungang Grottoes and Mount Heng) but funds for the huge city wall construction project come from municipal finances.[31] In all, these projects may total 50 billion RMB, a vast sum considering that Datong municipality's revenue amounted to just 14 billion RMB in 2010 (Honesty Outlook 2011). Revenue shortfalls have been made up with bank loans, land sales, the marketization of some public services and high ticket prices for the refurbished sites.[32] Geng has, reportedly, begun to run out of credit with local banks and a saying circulating in Datong officialdom during our interviews was that "Geng cannot get a loan" (*geng ban bu liao daikuan* 耿办不了贷款).[33]

6. XIAOYI: THE BENEFITS OF LOW CADRE TURNOVER AND LEADERSHIP CONTINUITY

> "You know, the worst thing for a local government is frequent change in the leading group and planning. In Xiaoyi, the mayor will usually become the next Party secretary so the top leaders of the two terms are old partners meaning there is good continuity."[34]

In contrast to Datong's one-man-show approach to economic transformation, Xiaoyi's greening growth strategy has built up gradually with guidance from a strong and locally rooted leadership group. Like Datong, Xiaoyi is a resource-based economy in the midst of transformation. Xiaoyi is one of China's 50 largest coal bases and home to about 20 percent of China's total bauxite reserves.[35] Coal coking is especially important to the local economy and drove a period of fast-paced GDP growth in the late 1990s and early 2000s. Over the past two decades the local leadership's priorities have evolved in step with those of the central government, from an early more conventionally developmental focus on maximizing economic growth and building roads and physical infrastructure, to the current leadership's green growth strategy based on reducing the proportion of heavy industry and cleaning up the industry-ravaged local environment. City leaders' efforts in establishing a "Xiaoyi paradigm" have won the locality plaudits from the upper levels of government and a long list of distinctions. A crucial element of Xiaoyi's success has been effective *guanxi*-building, especially with provincial and central governments as well as with local industry. Titles awarded by upper levels of government have brought in financial and advisory support and, working closely with local industry, Xiaoyi leaders have effectively induced the private sector to share the financial burden of greening growth. A high degree of continuity in the *lingdao banzi* has contributed to Xiaoyi's strong government and business networks because valuable connections are maintained and strengthened over time rather than discarded with each changing of the guard.

6.1. Leadership Cycles

The continuity and rootedness of Xiaoyi's *lingdao banzi* sets it apart from most Chinese counties. Over the past two decades, the usual pattern in Xiaoyi has been for leaders to move up from mayor to Party secretary, holding each position for a considerable length of time. In Xiaoyi, Party secretaries stay an average of 8.3 years while recent mayors have stayed 6.2 years, much longer than most county leaders in China. A former deputy mayor told us that leadership continuity had contributed stability to local planning and development:

> Every new period the leaders are doing better. The mayor and Party secretary in the 1990s built roads to connect Xiaoyi with the outside. The next leaders started industrial restructuring. The current Party secretary has placed greater emphasis on equitable growth and environment as well. Each period's achievements serve as inspiration for the next.[36]

The city's open culture is credited with encouraging leaders from elsewhere to put down roots in Xiaoyi: "Most of the secretaries, mayors and CCP Organization Department heads are from outside. But they all settle down here."[37] Xiaoyi's leaders also stress the importance of different arms of government pulling on the same string and, to that end, the preceding Party secretary, Li Liangsen, built a modest office building in order to bring all major departments under the same roof. Interviews conducted with various government departments in Xiaoyi suggested to us the existence of a fairly cooperative culture between departments as officials often called each other for documents and figures and are well informed about activities of other departments. Anecdotes about the leaders' attention to details also suggest that they have an uncommon regard for their locality. Current Party secretary Zhang Xuguang, who has been in Xiaoyi's *lingdao banzi* for ten years already, surprised the local EPB head when he asked him, out of the blue, what formula the EPB employs to calculate chemical oxygen demand (COD) reduction.[38]

Low cadre turnover has also helped to cement a long-term, goal-oriented culture among leaders and key departments. In 2002, for example, the Xiaoyi leadership first fixed on the goal of attaining a central-level experimental city designation (*shidian chengshi* 试点城市) to aid their economic transformation. Local leaders were dispatched to learn from experimental cities in the Northeast in preparation for their application.[39] Finally, in 2009, Xiaoyi was named a *Resource-Exhausted Transformation Experimental City*, one of only 44 nationwide and the only Shanxi locality found on the list. The title is not just a bright spot on the list of local leaders' accomplishments, it has also brought the city 200 million RMB in central government funding as well as policy support from National Development and Reform Commission (NDRC).[40] In a similar fashion, the Xiaoyi leadership decided in 2005 to become one of China's *100 Strongest Counties*, an award based on measures of economic competitiveness, wealth, environmental quality, and resident satisfaction. In 2006, Xiaoyi first entered the list at No. 93 and steadily climbed to No. 66 in 2011; it is the only Shanxi locality to be listed. These awards also have value in terms of helping local leaders establish and strengthen links to upper levels of government. Leaders from central government departments including NDRC have made official visits to Xiaoyi and the Shanxi government holds up Xiaoyi as a model for others to follow. Indeed, one of the province's aims for 12th FYP is to "create 10 or more such strong counties as Xiaoyi."[41]

The good connections Xiaoyi leaders maintain with local industry have also proven to be an important resource in developing and funding greening growth initiatives. The local government has been particularly effective in drawing local coal-based enterprises into economic "transformation projects" (*zhuanxing xiangmu* 转型项目). Between 2006 and 2010, Xiaoyi launched 362 such transformation projects worth a total of 99 billion RMB (Xiaoyi Government Work Report 2011).[42] The city's showcase transformation project is an LED light production facility, which is projected to generate approximately 1.7 billion RMB in tax revenue, about one-third of the city's annual intake.[43] While this is formally a private-sector initiative, Xiaoyi leaders were the initiators behind the scenes: "The government's role was to match the money with the technology. The two stakeholders of this project are Jinyan [a local, privately-owned coking enterprise] who provided the funding and Doctor Wu Yongan, a physicist from Stanford University, who provided the technology."[44] Local leaders' effort to induce the private sector to shoulder greening growth projects suggests a high degree of local state corporatism. Unusually, local coal enterprises, many of which are privately owned, were allocated soft targets for investment in transformation projects (Xiaoyi Government Work Report 2011:20). In Xiaoyi, the leadership has also effectively "bundled" coal restructuring with the goal of developing non-coal industries by providing incentives for former coal bosses to open greener businesses in the second and third sectors.[45] For instance, with government backing, a former mine owner whose enterprise was closed brought a Walmart outlet to Xiaoyi and has a new business of marketing agricultural products.[46]

6.2. Greening Growth

Transformation of the local economy had been on the minds of Xiaoyi leaders since the early 2000s, but it was a stern warning from the province on pollution in 2006 that was the catalyst for a major clean-up of the coal industry. In 2002, Xiaoyi began its first effort to reduce pollution generated by coking enterprises, many of which were operating illegally. The crackdown was successful in closing down scores of smaller coking enterprises but, in the years following, many illegal operations resurfaced.[47] A 2006 inspection by the provincial EPB office found that of Xiaoyi's 47 coking operations, only nine had an approved environmental impact report.[48] The newly appointed provincial EPB head decided to make an example of Xiaoyi and in August 2006, Shanxi EPB temporarily revoked Xiaoyi EPB's rights to conduct environmental evaluations (*quyu xianpi* 区域限批), effectively a block on the city's ability to approve new industrial projects.[49] This was the first time Shanxi EPB had made use of this severe penalty and the ensuing media glare put Xiaoyi leaders under extreme pressure. The leaders responded with a broad and costly crackdown on local industry, which focused on coking but extended to power, coal mining, and chemicals. Upon lifting of the *quyu xianpi* order in June 2007, Party secretary Zhang said: "We eliminated old production facilities and lost more than 20 billion RMB of assets and 450 million RMB in government revenue. But in the long run, without this pain, Xiaoyi city would not have taken the next step towards better and fast development."[50] Afterwards, the local government forged ahead with coal restructuring. In 2008, Xiaoyi's 400 coal mines were reorganized as 13 large-scale coal mines. 12 large enterprises now dominate the coking industry and small, polluting operations have not returned, partly due to strict environmental regulations requiring every enterprise to have waste water treatment, flue gas desulphurization, and smoke and dust removal facilities.[51]

The motto for Xiaoyi's approach to greening growth, "Taking coal as the base, diversifying development" (*yi mei wei zhu, duoyuan fazhan* 以煤为主，多元发展), reflects the leadership's measured approach to greening growth as compared to Datong. As one interviewee told us, the aim is to have "longer value chains for the first sector, less emissions from the second and more investment in the third."[52] In agriculture, Xiaoyi government is supporting the farming and processing sides of the local walnut business, both to boost agricultural income and because walnut trees are especially good at carbon sequestration. In the second sector, the leadership's efforts to clean coal hinge on scaling up coal mining and coking operations and promoting the establishment of "circular economies" (*xunhuan jingji* 循环经济) in which industrial waste is used as inputs for other processes.[53] Xiaoyi is also actively promoting local manufacturing. The large-scale LED project described above has received strong backing from local government because LED lights use much less energy than conventional bulbs.[54] Newly established battery-powered scooter manufacturing enterprises also receive government support. Finally, expanding the proportion of the third sector from the current 36 percent to 45 percent in 2015 is an official target, though a soft one, in the 12th FYP and logistics, tourism, and finance are the focus industries.[55] Some coal bosses whose operations were closed after 2006 have received government support for starting businesses in the third sector.

The media is fond of saying that Xiaoyi government took its "yellow card" on pollution and turned it into a "gold medal" on environmental protection.[56] At the end of 2010, Xiaoyi added to its list of distinctions when it was awarded the title of *National Green Model City* for the results of the government's 650 million RMB investment in "blue skies, clear water" (*lan tian bi shui* 蓝天碧水) projects. In the 11th FYP period, Xiaoyi dramatically raised its forest coverage rate from 19 percent to 35 percent and urban area green space from 31 percent to 43 percent.[57] Xiaoyi's greening rate ranks first in the province.[58] Elimination of highly polluting coking operations and the replacement of coal-fired boilers with district heating facilities has also reduced air pollution. The number of days with air quality of Level Two or higher increased from 97 in 2005 to 354 in 2010 (Xiaoyi Government Work Report 2011:6). In the wake of Xiaoyi's "yellow card" from Shanxi EPB, local leaders also worked to strengthen local environmental agencies. Most importantly, the administrative rank of EPB's Environment Monitoring Team was upgraded to division (*ke* 科) level to increase their power vis-à-vis industry and thereby ease rule enforcement. Xiaoyi's Monitoring Team is the only one to have such a high rank in Shanxi province. The local EPB also secured additional support from the city in the provision of personnel and facilities.[59] Similarly, the Xiaoyi Gardening Bureau also receives an unusually high degree of support from Xiaoyi government. Their annual budget of 100 million RMB for greening work is equal to the combined total of all ten other counties in Lüliang municipality.[60]

Xiaoyi leaders are deservedly proud of their accomplishments in recent years but there are also factors inhibiting the city's green rise. First and foremost, to some degree, the local government seems to be paying lip-service to development of a robust service sector. In a manner consistent with the model of green growth policy prioritization presented above, local leaders seem to be putting off some initiatives with low *zhengji* value and a long time to maturity. As one informant put it,

For local government, the second sector is the most important because it contributes the most to government revenue, which is a key indicator in the evaluation of a county or municipality. Only when the percentage of non-coal industry becomes an indicator on these evaluations will local governments have sufficient incentive to really develop the third sector.[61]

Similarly, for all their emphasis on building the foundations of "quality" growth, Xiaoyi leaders take a fairly conventional view of the desired speed of economic growth. Despite the fact that Xiaoyi residents' average income is high by Shanxi standards, Xiaoyi's GDP average annual growth targets for the 12th FYP are at a minimum 20 percent and at a maximum 25 percent—well above the 7 percent national target.[62]

7. Discussion and Conclusion

In the discussion section, we return to our research question: Does high turnover of cadres in the *lingdao banzi* help or hinder China's green rise? The case studies suggest that cadre turnover is something of a drag on Beijing's green growth ambitions. From the perspective of the center, the function of institutionalized cadre turnover is to prevent leaders from falling captive to local interests and keep them responsive to decrees from Beijing by limiting the length of their stays in each post. Previous work suggests that high turnover can encourage compliance with readily measurable, on–off policies such as investment controls. But the complex, uncertain policies on the greening growth agenda may not be served well by short-stayers. Since Beijing has provided only the first brushstrokes of its green vision leaving local leaders to fill in the image, localities with a high implementation burden under the new green plans will need effective and locally rooted leaders with a long-term vision. The findings presented here suggest that the many difficult and costly aspects of greening growth in places like Datong and Xiaoyi—including reducing pollution, restructuring traditional industries, and developing new ones—demands that leaders have an expert's grasp of the local economy as well as the political and financial capital to administer bitter pills. Such qualities might only rarely describe the tenure of local leaders who cycle in and out of the leadership group at three- or four-year intervals. On the basis of our casework, we discuss the impact of high cadre turnover on three factors that seem crucial to successful greening growth initiatives: time, strong leadership, and money.

First, greening growth is a gradual process that is not easily compressed into the term of a typical leading cadre at the municipality or county level. In Datong, the need to change lanes from a coal-dependent industrial structure to a diversified, greener economy had long been acknowledged, but a string of uninspiring leaders effectively passed the buck until Geng Yanbo's arrival on the scene in 2008. Yet, even a maverick *gexing guanyuan* like Geng operates within the constraints of a system that expects him to move on after five years. The immense time pressure imposed by Mayor Geng's expected exit in 2013 accounts for his headlong approach to the transformation of Datong, the risks of which are considerable. Xiaoyi's approach to greening growth is a study in contrasts. In Xiaoyi, a high degree of continuity in the local leadership group has helped to make long-term planning a habit of local rule. Xiaoyi leaders began to focus on transforming the local economy as early as 2002 and have developed and implemented their plans in stepwise fashion over the last decade. Local leaders' good working relationships with each other and with local industry also help to explain why Xiaoyi was able to respond quickly and decisively to the yellow card from the Shanxi EPB in 2006 in sharp contrast to other localities under similar degrees of external pressure.[63] The leadership group has also made very effective use of the relationships it has built over time with upper levels of government and especially with local industry. These precious *guanxi* resources are effectively preserved over time since long-serving mayors typically become long-serving Party secretaries in Xiaoyi.

Second, a strong, sustained leadership is needed to build the foundations of greener growth, not least since many associated changes impose high costs on local businesses and local employment. Leaders in both Xiaoyi and Datong led costly restructuring initiatives that succeeded in pushing the heaviest polluters out of business. By contrast, in many other Shanxi localities facing similar problems, such as Linfen (see footnote at the bottom of this page), weak leaders continue to employ delaying tactics. While Geng's strong-man leadership has provided an answer to the problem of Datong's coal dependence, his risky tourism-focused strategy is based on his own somewhat idiosyncratic vision of the city's latent strengths. The plans in Xiaoyi are comparatively less bold and this likely reflects the collective input of the leadership group as well as local industry. Indeed, while both models of greening growth examined in this study can be seen as successful (or at least promising) cases, Xiaoyi's seems the more readily sustainable because it is not pinned to the ambitions of a single person but carried forward by a more collectivist leadership. In Datong, it remains to be seen whether the energy and vision Geng has brought to the city will survive his departure.

Third, changing lanes is expensive and here, too, our cases point to the benefits of low cadre turnover in paying for greener growth. Xiaoyi's farsighted, collectivist leaders make very effective fundraisers. Their focus on winning national awards and experimental city designations has brought in major funding as well as planning support from upper levels. They have also used their *guanxi* ties to induce local businesses to share the financial burden of reducing Xiaoyi's coal dependence. Xiaoyi leaders effectively "bundled" coal restructuring with the goal of developing non-coal industries by providing incentives for bosses whose enterprises were eliminated in the 2006 industry clean-up to start greener businesses in the second and third sectors. Xiaoyi leaders have also taken the unusual step of giving surviving coal enterprises soft targets for investment in transformation projects. Leadership continuity has probably contributed to the leaders' success in securing investment because investors can be confident that plans will not shift radically with personnel changes in the leadership group. By contrast, Datong's greening growth projects have been paid for with whatever Mayor Geng could cobble together, principally bank loans and municipal land sales. While Geng has had some success in attracting central SOEs to invest in green projects, his approach relies much less on local business burden-sharing, partly because, as a new arrival in 2008, he did not have connections to local industry on which to draw.

How might China's policymakers mitigate the downsides of high cadre turnover suggested by these case studies? While this is a difficult question deserving of a paper on it, our analysis suggests the following preliminary policy implications. First, rapid cadre turnover encourages cadres to sideline longer-term green growth initiatives and gather only the low-hanging fruits most helpful for their career advancement. A step forward in the effort to lengthen cadres' time horizon is the central government's 2006 "Interim Provisions on the Tenure of Leading Party and Government Cadres," which states that leading cadres should serve out their five-year terms in full. Second, China's leaders could consider modifying the planning and cadre evaluation systems. While many see the hardening of environmental targets in the 11th and 12th FYPs as a promising step in terms of incentivizing leading cadres to prioritize environmental protection and greening industry, these new incentives are probably not, on their own, adequate to induce local leaders to initiate the kinds of broad based transformation described in this chapter. Although the new plans are intended to promote greening growth, local leaders have myriad ways of satisfying the short-term goals contained in these plans while staving off needed costly and difficult change. To curtail this short-termism and encourage leaders to take the long view, one solution might be to move away from the current focus on quantitative, discrete measures in the planning and cadre evaluation systems to some form of comprehensive, qualitative evaluations of greening growth. Efforts to increase the after-tenure accountability of cadres might also serve to mitigate the problems associated with "tenure rush."

Of course, cadre turnover is just one aspect of a complex picture. Our case studies of greening growth initiatives in the two coal-dependent localities touched on a number of factors emphasized in previous work including: the availability of funding from the central government, central–local interest divergence, inherited industrial structures, bureaucratic coherence and implementation capacity, leaders' personalities, shared values among cadres in the *lingdao banzi*, cadre succession modes, and the cadre evaluation system. Alongside these influences, our cases suggest that the frequent turnover of local leaders helps to explain why comprehensive green growth projects are often sidelined. In Datong, with the exception of the current mayor, local leaders are flitted through their terms regardless of whether or not they have had enough opportunity to initiate comprehensive industrial transformation efforts. This changed with the arrival of Geng Yanbo, who initiated a large-scale tourist-based development strategy. While there is reason to be cautiously hopeful about this approach, Geng's personal time horizon has added risk and expense to Datong's bid to change lanes. By contrast, in Xiaoyi, a string of long-serving leaders in the *lingdao banzi* contributed stability to local strategic projects and drew funding from upper government and local businesses. Serving longer terms also helped Xiaoyi's leaders to develop and maintain *guanxi* resources built over time and ensured that developed strategies were realistic and fit Xiaoyi's untapped potential.

The analysis is exploratory in nature, relying on two in-depth case studies framed by a set of loose conjectures about how cadre turnover might affect local implementation of Beijing's green growth mandate. As such, we have identified dynamics with potentially broader significance but additional research is needed to assess the generalizability of these findings. Quantitative analysis could contribute to a better understanding of the pros and cons of frequent post-shuffling among local government cadres and assess how much of the variance in green growth implementation is accounted for by cadre turnover. While reliable data are available for cadre turnover at municipal and provincial levels, quantitative measures of "green growth" initiatives would need to be collected through survey analysis or through the development of a comprehensive environmental implementation index.

NOTES

1. Three SEIs—alternative energy, clean energy vehicles, and clean energy technology sectors—align with the broader goal of conserving resources. The other four SEIs—biotechnology, new materials, next-generation IT, and high-end equipment manufacturing—align with the aim of moving up the value chain. See Casey and Koleski (2011:18).
2. The main government regulations related to renewable energy and energy conservation are a Renewable Energy Law (2006), a revised Energy Conservation Law (2007), and a Circular Economy Promotion Law (2009).
3. For instance, officially, Inner Mongolia reported they had met and exceeded energy intensity reduction targets with 23 percent during the 11th FYP period. The accuracy of these figures is, however, doubtful since, in September 2010, three months before the end of the 11th FYP period, leading officials in Inner Mongolia indicated that the province was far from meeting the target (Interview 65, Leading Provincial Official in Inner Mongolia, September 2010).
4. For a detailed description of how the cadre responsibility and evaluation systems promote environmental governance, see Heberer and Senz (2011). For a detailed description of distinctive characteristics of environmental policies, see Eaton and Kostka (2014).
5. The post-switching of leading cadres' (*lingdao ganbu* 领导干部) takes place via two systems within the cadre management bureaucracy, both of which are managed ultimately by the CCP Organization Department. The appointment system handles promotion and demotion decisions for leading cadres while the cadre rotation system (*ganbu jiaoliu zhidu* 干部交流制度) applies to cadre flows between positions of equal rank.
6. Other salient factors might include: the relative importance of factional ties and office-buying as an advancement strategy; implementation capacities;

available funding; measurability and verifiability of targets; leaders' previous working background, personal preferences, and beliefs.
7. Interview 23, Leading Government Official at District Development and Reform Commission, September 27, 2011.
8. Interview 34, Leading Government Official in Datong, September 30, 2011.
9. Interview 37, Manager of Private Coal Mine in Lingshi, July 16, 2010.
10. Interview 34, Leading Government Official in Datong, September 30, 2011.
11. Ibid.
12. Nailheads are households that refuse to relocate when the land is requisitioned.
13. Interview 37, Manager of Private Coal Mine in Lingshi, July 16, 2010.
14. Interview 20, General Manager of Real Estate Company in Datong, July 6, 2010.
15. Interview 37, Manager of Private Coal Mine in Lingshi, July 16, 2010.
16. Interview 2, Government Official in Datong, September 1, 2011.
17. Interview 26, County-Level Government Official in Development and Reform Commission, September 28, 2011.
18. Interview 34, Leading Government Official in Datong, September 30, 2011.
19. Interview 20, Government Official in Development and Reform Commission in Datong, September 26, 2011.
20. At the time of Geng's speech the most important national policy on coal was NDRC's 2007 *11th Five Year Plan for Coal Industry Development*. The plan calls for large coal bases such as Datong to promote industrial consolidation around large enterprises and raise industry concentration rates.
21. Datong Development and Reform Commission, "Guanyu datong shi 2010 nian guomin jingji he shehui fazhan jihua zhi hang qingkuang yu 2011 nian guomin jingji he shehui fazhan jihua de baogao" ("On Datong's 2010 economic and social development implementation plan and the 2011 report on the national plan for economic and social development"), June 16, 2011, 4.
22. Interview 34, Leading Government Official in Datong, September 30, 2011.
23. Datong county is one of the seven counties and four areas under Datong municipality.
24. Interview 29, County Government Official in Development and Reform Commission in Datong, September 29, 2011.
25. Datong County Economics and Business Information Bureau, "Gongzuo huibao" ("Work Report"), 14.
26. Interview 34, Leading Government Official in Datong, September 30, 2011.
27. Interview No. 22, Government Official in Economic Commission in Datong, September 23, 2011.
28. Interview 20, Government Official in Development and Reform Commission in Datong, September 26, 2011.
29. Datong Development and Reform Commission, "Guanyu datong shi 2010 nian guomin jingji he shehui fazhan jihua zhi hang qingkuang yu 2011 nian guomin jingji he shehui fazhan jihua de baogao", 7.
30. For instance, oil giant China National Offshore Oil Corporation (CNOOC) and Datong Coal Mining Group have jointly invested in a 30 million RMB natural gas project near Datong.
31. Interview 34, Leading Government Official in Datong, September 30, 2011.
32. Ticket prices have reportedly skyrocketed since 2008. The total cost of entry to Datong's four major attractions—Huayan and Shanhau temples, the

Yungang Grottoes and Xuankong "hanging" temple—is US$65 per person making it quite unaffordable to most locals, see Ian Johnson, "China's Glorious New Past" *NYR Blog*, June 1, 2011, available at: http://www.nybooks.com/blogs/nyrblog/2011/jun/01/chinas-glorious-new-past/.
33. Interview 34, Leading Government Official in Datong, September 30, 2011.
34. Interview 15, Retired Government Officials in Xiaoyi, September 21, 2011.
35. Xiaoyi City People's Government, "Chuangzao sheng ji huanbao mofan chengshi gongzuo baogao zhonggong xiaoyi shiwei, xiaoyi shi renmin zhengfu" ("Work report on creation of a provincial-level environmental protection model city 'CPC Xiaoyi City' "), 2.
36. Interview 15, Retired Government Officials in Xiaoyi, September 21, 2011.
37. Ibid. While it falls outside the scope of this research, the attractiveness of Xiaoyi as a place to live does seem to have contributed to its success. Living environment is often cited as an important factor in luring investors, it may also be important in attracting and retaining able leaders.
38. Interview 12, Head of Environmental Protection Bureau in Xiaoyi, September 20, 2011.
39. Interview 11, Government Official in Development and Reform Commission in Xiaoyi, September 26, 2011.
40. Ibid.
41. "Zhongong xiaoyi shiwei xiaoyi shi renmin zhengfu guanyu shenqing pizhun 'xiaoyi shi zonghe peitao gaige shiyan zongti fangan' de baogao", ("Report by Xiaoyi City Party Committee and Xiaoyi People's Government on approval of 'Xiaoyi City Comprehensive Reform Pilot Program' application"), 2.
42. Xiaoyi leaders pride themselves on their skills in attracting investment for economic "transformation projects" (*zhuanxing xiangmu* 转型项目). Long-time head of Xiaoyi DRC, Ren Huachao, notes as one of his personal achievements having played a lead role in attracting nine billion RMB worth of *zhuanxing* investment from French, German and Chinese companies.
43. "Xiaoyi: bian 'wuran huangpai' wei 'huanbao jinpai' ", ("Xiaoyi is changing 'pollution yellow card' to the 'green gold medal' "), August 14, 2011, available at: http://news.sxrtv.com/shtml/0/572/content572402.shtml?pid=147&CatalogNumber=sxxwzhgb01&ProgramID=572402
44. Interview 16, Manager of Private Coal Coking Enterprise in Xiaoyi, September 22, 2011.
45. On policy and interest bundling, see Kostka and Hobbs (2012).
46. Interview 15, Retired Government Officials in Xiaoyi city (previously: Head of DRC and Director of People's Congress), September 21, 2011.
47. Xiaoyi City People's Government, "Chuangzao sheng ji huanbao mofan chengshi gongzuo baogao zhonggong xiaoyi shiwei, xiaoyi shi renmin zhengfu" ("Work report on creation of a provincial-level environmental protection model city 'CPC Xiaoyi City' "), 3.
48. "Zanting Xiaoyi huanbao shenpi quan" ("Suspension of environmental approval rights in Xiaoyi"), *Shanxi xinwen wang*, (*Shanxi News Net*), September 8, 2006, available at: http://zqb.cyol.com/content/2006-09/09/content_1505526.htm.
49. Interview 8, Shanxi Provincial Head of Environmental Protection Bureau in Taiyuan, June 30, 2010.
50. "Xiaoyi: bian 'wuran huangpai' wei 'huanbao jinpai' ", ("Xiaoyi is changing 'pollution yellow card' to the 'green gold medal' ").

51. Interview 16, Manager of Private Coal Coking Enterprise in Xiaoyi, September 22, 2011.
52. Interview 10, Government Official in Xiaoyi, September 19, 2011.
53. The 12th FYP envisions 280 million RMB worth of circular economy projects ("Report by Xiaoyi City Party Committee and Xiaoyi People's Government on approval of 'Xiaoyi City Comprehensive Reform Pilot Program' application", 6). Coal waste recycling projects (e.g., using fly ash to build bricks) are typically paid for by the enterprises themselves and government support comes in the form of tax exemptions.
54. Interview 16, Manager of Private Coal Coking Enterprise in Xiaoyi, September 22, 2011.
55. Xiaoyi People's Government, "Xiaoyi shi renmin jingji he shehui fazhan di shi erg e wu nian guihua gangyao" ("Outline of Xiaoyi's Economic and Social Development 12th Five Year Plan"), June 2011, 57–61.
56. E.g., "Xiaoyi: bian 'wuran huangpai' wei 'huanbao jinpai' ".
57. "Outline of Xiaoyi's Economic and Social Development 12th Five Year Plan", 12.
58. Interview 19, Deputy Director of the Forestry Bureau in Xiaoyi, September 23, 2011.
59. Interview 12, Head of Environmental Protection Bureau in Xiaoyi, September 20, 2011.
60. Interview 18, Director of the Garden Bureau in Xiaoyi, September 23, 2011.
61. Interview 17, Government Official in Development and Reform Commission in Xiaoyi, September 21, 2011.
62. Xiaoyi's GDP per capita was 45,538 RMB in 2009, twice as high as the provincial GDP per capita level averaging at 21,506 RMB in 2009 (Shanxi 2010 Statistical Yearbook, Tables 19.3, 636–638).
63. Another Shanxi municipality, Linfen, offers a striking contrast. It ranked as one of the most polluted cities among China's 113 key cities inspected by the national Ministry of Environmental Protection in 2005. Yet, despite extreme pressure from provincial and national governments, Linfen did not fulfill most of the greening growth targets in the 11th FYP. For instance, by summer 2010 Linfen had only achieved 60 percent of its total 11th FYP energy efficiency targets and provincial leaders complained about the lack of cooperation from Linfen's leaders. Part of the problem was that the municipality had no Party secretary for more than eight years, mainly because the frequent occurrence of mine accidents and work safety scandals customarily led to leaders' resignation.

Bibliography

Cai, Yongshun. 2004. "Irresponsible State: Local Cadres and Image-Building in China." *Journal of Communist Studies and Transition Politics* 20(4): 20–41.
Casey, Joseph and Katherine Koleski. 2011. "Backgrounder: China's 12th Five-Year Plan." *U.S.-China Economic & Security Review Commission* June 24, 1–19.
Datong Daily (Datong Ribao). 2008. "Acting Mayor Geng Yanbo." July 11. Accessed February 4, 2012.
Datong Government Work Report. 2011. "Datong Zhengfu gongzuo baogao" June 24, Datong City.

Eaton, Sarah and Genia Kostka. 2014. "Authoritarian Environmentalism Undermined? Local Leaders' Time Horizons and Environmental Policy Implementation." *The China Quarterly* 218: forthcoming.
Elizabeth C. Economy. 2004. *The River Runs Black the Environmental Challenge to China's Future*. Ithaca, NY: Cornell University Press.
Fewsmith, Joseph. 2006. "Promotion of Qiu He Raises Questions about Direction of Reform." *China Leadership Monitor* no. 17: 1–8.
Fewsmith, Joseph. 2010. "Bo Xilai Takes on Organized Crime." *China Leadership Monitor* no. 32: 1–6.
Fu Peiyi, Ge Yonghui, Ma Chao, Jia Xiuming, Shan Xinjian, Li Fangfang and Zhang Xiaoke. 2010. "A Study of Land Subsidence by Radar Remote Sensing at Datong Jurassic and Carboniferous Period Coalfield." *Proceedings of the 2010 3rd International Congress on Image and Signal Processing*, pp. 1–9.
Guo, Gang. 2009. "China's Local Political Budget Cycles." *American Journal of Political Science* 53(3): 621–632.
Harrison, Tom and Genia Kostka. 2014. "Balancing Priorities, Aligning Interests: Developing Mitigation Capacity in China and India." *Comparative Political Studies* 47(3–4): forthcoming.
Honesty Outlook. 2011. "Datong shi zhang geng yanbo 15 nian de chengjian zhengzhi" ("The 15 years of construction politics of Datong's Mayor Geng Yanbo"), *Lianzheng liaowang (Honesty Outlook)*, September 28. Accessed May 14, 2013. http://news.sina.com.cn/c/sd/2011-09-28/125923231364.shtml.
Huang Jingwen. 2011. "China Prepares to End GDP Obsession." *China Daily* March 07, Accessed May 13, 2013. http://www.chinadaily.com.cn/china/2011npc/2011-03/06/content_12124349.htm.
International Energy Agency. 2009. *Cleaner Coal in China*. http://www.iea.org/textbase/nppdf/free/2009/coal_china2009.pdf
Kostka, Genia and William Hobbs. 2012. "Local Energy Efficiency Policy Implementation in China: Bridging the Gap between National Priorities and Local Interests." *The China Quarterly* 211: 765–785.
Kostka, Genia and William Hobbs. 2013. "Embedded Interests and the Managerial local State: The Political Economy of Methanol Fuel-Switching in China." *Journal of Contemporary China* 22(80): 204–218.
Lieberthal, Kenneth and Michel Oksenberg. 1988. *Policy-Making in China: Leaders, Structures, and Processes*. Princeton, N.J.: Princeton University Press.
Ma, Jingbo and Zhang Zhihong. 2006. "Datong zhai diao daqi wuran 'hei maozi' " ("Datong is taking off the air pollution 'black hat' "), *Shanxi Ribao (Shanxi Daily)*, August 29, Accessed May 13, 2013. http://news.sina.com.cn/s/2006-08-29/03459874550s.shtml.
O'Brien, Kelvin J. and Lianjiang Li. 1999. "Selective Policy Implementation in Rural China." *Comparative Politics* 31(2): 167–186.
Ran, Ran. 2013. "Perverse Incentive Structure and Policy Implementation Gap in China's Local Environmental Politics." *Journal of Environmental Policy and Planning* 15(1): 17–39.
Shanxi Statistical Bureau. 2011. Shanxi 2010 Statistical Yearbook (Shanxi 2010 tongji nianjian). Beijing: China Statistical Press.
The Beijing News. 2011. "Shi zhang 'deng bu qi', ze qiang chai ting buxia" ("Mayor will not Stop Demolitions"), *Xin Jing Bao (The Beijing News)*, May 22, Accessed February 4, 2012. http://news.xinhuanet.com/comments/2011-05/22/c_121443752.htm.

Thomas, Heberer and Anja Senz. 2011. "Streamlining Local Behaviour Through Communication, Incentives and Control: A Case Study of Local Environmental Policies in China." *Journal of Current Chinese Affairs* 40(3): 77–112.

Van Rooij, Benjamin. 2006. "Implementation of Chinese Environmental Law: Regular Enforcement and Political Campaigns." *Development and Change* 37(1): 57–74.

Wang, Yan. 2010. "Datong qunian kongqi wuran zhishu jiangfu quan sheng di yi" ("Last year Datong's air pollution decreased the most in Shanxi province"), *Shanxi Jingji Ribao* (Shanxi Economic Daily), 2nd Version, February 6.

Xiaoyi Government. 2011. Xiaoyi Government Work Report ("Zhengfu gongzuo baogao"), June 24, 2011, pp.1–35.

Zhong, Yang. 2003. *Local Government and Politics in China—Challenges from Below*. Armonk, New York: M.E. Sharpe.

Chapter 4

Incentive Structures and Compatible Development in a Chinese Local State

Jianguo Chen

1. Introduction

Environmental protection and economic growth seem to be two contradictory concepts. Exploiting natural resources in order to improve living standards, according to some observers, will inevitably cause environmental damage, which, ironically, would eventually threaten our health and even our survival. On the other hand, policies aiming to reduce pollution and improve efficiency may drive up the costs for economic growth and therefore discourage industrial competitiveness in international markets.

This understanding leads to two contradictory policymaking models. One side argues for the legitimacy of growth determinism, believing that environmental protection needs to give way to growth until the living standards have improved substantially. Contrary to this "pollute first, fix later" argument, others advocate for a zero-growth theory that puts environmental protection above the need for economic growth. Could there be a third way that allows us to pursue both economic growth and environmental protection simultaneously?

Research on this third line of thinking abounds but applying it to China's reality seems challenging. Is it possible to coordinate environment and growth in a country highly dependent on natural resources and heavy industries? If so, what lessons can we learn from the practices in some areas in China? The experiences in China will have significant implications not only for China's sustainable development in the future but also for other developing countries that face the similar dilemma.

This chapter is organized as follows. The first section reviews literature of the dispute over the relationship between environmental protection and economic growth. The second section puts forward an analytical framework of compatible development. The third section applies this framework into Shanxi province and analyzes its environmental governance strategy that has been relatively successful in coordinating environmental protection and economic growth. An empirical analysis in Section 4 gives a more precise assessment of the effect of this model in Shanxi. The conclusion section summarizes the policy implications for China's environmental governance.

2. Debates over Environment and Growth

The longstanding debate over the relationship between environmental protection and economic growth has taken place at two levels of analysis. At the macro-level, some scholars argue that economic growth threats environment and advocate the slowdown of economic growth for environmental protection. Prominent scholars such as Georgescu-Roegen (1971), Ehrlich and Holdren (1971), Georgescu-Roegen (1971), and Meadows (1972), based on the principle of energy conservation, argued that unlimited economic growth must cease because continuous growth in the context of finite resources is unlikely to support current levels of prosperity indefinitely. Instead they advocate a zero-growth model in which all economic activities and policies are oriented toward achieving a state of equilibrium in order to avoid a "boom/bust" cycle that is inherent in the conventional continuous growth model (Nicholas 1971). Along with the various studies following this line of thinking, the most influential one is *The Limits to Growth* (Meadows, Meadows, Randers, and Behrens, 1972) by the Club of Rome. The authors of that study argue that human societies would not support more outputs such as foods, services, and other consumer items when industries decline. The cease of growth in these sectors will put an end to population growth as well. Jansson, Hammer, Folke, and Costanza (1994) argued that degradation of the resource base would eventually put economic activities themselves at risk. To save the environment and even economic activity itself, economic growth must cease and give way to a steady-state economy, thus zero growth (128–294).

The other side of the debate argues instead that economic growth could go hand in hand with environmental improvement. Some researchers argued that higher incomes increase the demand for less material-intensive goods and services; at the same time higher incomes bring about an increased demand for environmental protection measures. As Beckerman famously claimed, "the strong correlation between incomes, and the extent to which environmental protection measures are adopted, demonstrates that in the longer run the surest way to improve your environment is to become rich" (1992:495). In an analysis of the changes of the amount of pollutant emission in 66 countries, Grossman and Kreuger concluded that there would be an inverted-U relationship between the amount of pollutant emission and per capita gross national product. That is to say, the pollutant emission will first increase and then decrease along with the increase of per capita income, a phenomenon known as the "environmental Kuznets curve" (Grossman and Krueger 1995).

At the micro-level, the debate focuses on the relationship between environmental protection and economic competitiveness for certain regions, sectors, industries, or firms. On one side, scholars argue that since rigid environmental regulations and higher environmental standards will add costs and force certain regions, sectors, or firms to reduce their productivity, environmental protection is contradictory to economic competitiveness. Some studies found correlation between environmental regulation and productivity of certain industries or firms: the productivity of manufacturers under more rigid environmental regulation is often lower and also grows slower than those with less regulation. To solve this problem and to attract investment, governments compete to lower their environmental standards, thus creating a phenomenon called "race to the bottom" (Klevorick 1996).

On the other hand, Porter (1995) argued that it is possible to maintain environmental protection and improve competitiveness at the same time because more rigid environmental regulations would promote innovation and lower costs, thus increasing net benefits and manufacturers' competitiveness in international markets. Known as "Porter Hypothesis," this argument challenges the static model of the relationship between firms and economic development in the neoclassical economics and instead introduces a dynamic model.

Chinese scholars recently have introduced and applied these concepts and arguments to China's context (Zhang 2001; Cao 2004; Xiong 2005; Xing and Liu 2006; Pi 2010; Zhang and Zhou 2009). Most of these studies focus on the coordination between economic growth and environmental protection or technological innovation. Very few, however, examine the role of governments, institutional constraints, and incentive structures in environmental governance as well as the behavior of the actors such as regulators and firm managers. As Sun (2006) argues, the government has paid too much attention to economic performance as the key index in governmental performance evaluation and official selection and appointment, only aggravating the imbalance between environmental protection and economic growth. Therefore, environmental protection indicators need to be introduced into the evaluation system. Qi and Zhang (2007), however, caution about the applicability of "Green GDP" in the official evaluation system, arguing that the concept is inadequate as an indicator. In a separate study (Zhang and Qi 2010), they treat local governments as rational actors and analyze their dilemma between political promotion and financial constraints. They propose to optimize the current evaluation system in order to give local governments more incentives to pay more attention to environment. Similarly, Zhou (2009) and Wang (2008) examine the evaluation system in various regions and argue for better design of the evaluation system. These studies offer useful insights for this chapter to develop a new model of compatible development. However, these studies pay primary attention to the necessity and implications of evaluation system and incentive mechanisms in easing the tension between environmental protection and economic growth. Although they have developed some useful index and indicators, these studies fail to address how existing evaluation system and incentive structures can be utilized to promote the coordination of environmental protection and economic development.

This chapter aims to shed lights on this issue by putting forward a compatible model, which will focus on how relevant actors act within the institutional constraints of the existing incentive structure with positive outcomes. Shanxi province is chosen to test this model because Shanxi is argued to be one of the most polluted regions in China or even in the world. How the situation improved in this resource-dependent region will have important implications for our understanding of the relationship between environmental protection and economic development in China and beyond.

3. Compatible Development Model: Logic of the Harmonious Relationship between Environment and Economy

To understand the motivation for environmental protection, we need to first understand the motivation for pollution. Individuals pollute because this is the cheapest solution to dealing with certain actual problems in the process of economic growth such as costs. This suggests that in order to make individuals more willing to protect environment, we need to pay attention to the incentives of individuals in their economic activities. In China, firms are both producers and polluters. Governments are promoters of economic growth but also the protectors (regulators) of environment. To understand the relationship between environmental protection and economic growth, it is critical to understand the relationship between Chinese firms and governments, and the incentive structures that affect the relationship between governments and firms. The key to having a positive outcome is to ensure that governments at higher levels offer appropriate incentives for the governments at lower levels, and governments offer appropriate incentives to firms. Figure 4.1 illustrates the relationship among these actors in an incentive network that can produce positive outcomes.

In this network, a government at a higher level sets targets for a lower-level government to achieve balance between economic growth and environmental protection. The tasks are broken down and distributed among different departments at each level of government. Both the government and its respective departments are accountable for the targets assigned. The government at the higher level then assesses the performance of the government at the lower level based on a set of indicators and accordingly rewards or

Incentive Structures and Compatible Development

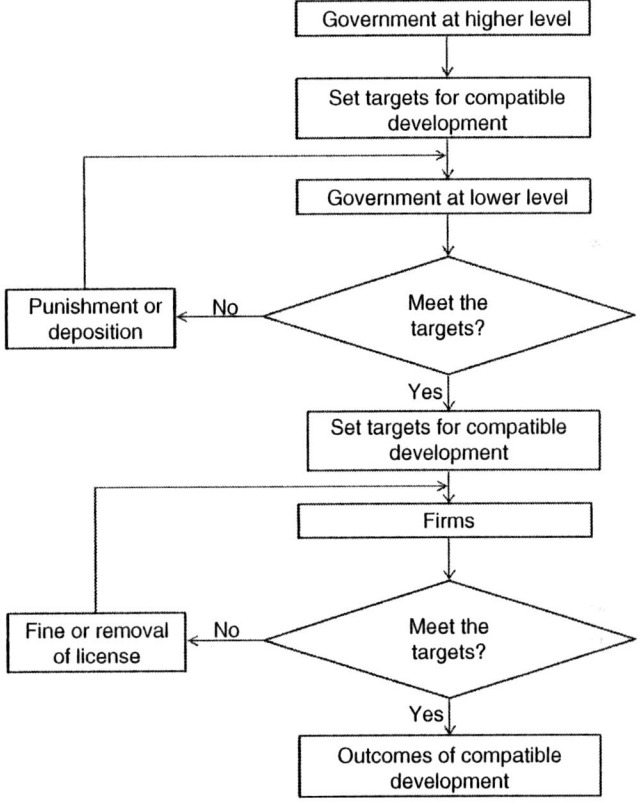

Figure 4.1 Incentive structures in a compatible development model

punishes. Likewise, the government at the lower level sets its own targets for the firms in its jurisdiction to follow, and then either rewards (for example, tax breaks, subsidies, special loans, etc.) or punishments (fine or removal of license) will follow once firms' performance is assessed.

With such a well-organized incentive network of accountability and assessment among governments at different levels and firms, firms will be motivated to meet the government's targets in order to seek more profits and improve their market competition. For the government, shutting down a few firms will not affect the economy of the entire region but may instead attract firms that pollute less. Through eliminating polluting firms and replacing them with more efficient firms, local governments can coordinate economic development and environmental protection.

It has become a consensus among many governments that it is necessary to take both economic development and environmental protection into policy consideration and many governments have implemented this approach in their evaluation system. But how well have they done in this transition from a GDP-dominated development strategy to a compatible development strategy? Shanxi province used to be one of the most polluted regions in China but in recent years has successfully realized the harmony between environmental protection and economic development. Its practices will offer insights for us to understand how the incentive structures have worked at the local level in China.

4. The Experience and Mechanism of Compatible Development in Shanxi Province

Shanxi was once the most polluted province in China. From 2003 to 2005, its three major cities, Linfen, Yangquan, and Datong, ranked the last three among 113 Chinese cities surveyed for air quality. Shanxi ranked number one in dust, solid waste, and hazardous waste emissions. By 2008, however, the urban air quality of eight cities and 35 counties in Shanxi achieved Grade II (the second best out of six grades, according to the National Air Quality Standard) for the first time, which indicated the significant progress in this aspect. The reduction of sulfur dioxide and chemical oxygen demand (COD) exceeded the task set by the central government, and reduction of SO_2 came out in front of the nation. At the same time, the gross domestic product (GDP) in Shanxi in 2009 reached 7,365.7 million *yuan*, up by 5.5 percent, suggesting that both environmental protection and economic growth were met. How did Shanxi achieve this balance?

Schumpeter (1934) suggested that entrepreneurs are critical for economic development because they are innovators that engage in "creative destruction." In market competition, firms make risky decisions to invest in new technology in order to seize new opportunities and gain market shares. Similarly, politicians with such a spirit seek innovations in order to enhance their political capital and influence. Shanxi's transition has been closely associated with political entrepreneurs represented by Liu Xiangdong, the head of the provincial environmental protection bureau, who in his inaugural speech (2006) promised the folks in this province to "see the sunshine in daytime and moonlight at night, and clear water in rivers." Liu believed that the problem with the previous system was the conflict of interests and fragmented power among different departments. To break the inertia of the old system, Liu departed from the usual routine of officialdom and wrote open letters to the heads of governments at the lower levels and demanded them to take direct responsibility of local environmental governance. He wittily explained that, to catch a mouse, you don't do it by yourself. The better way to do it is to get the cat owner—referring to local officials—to do the job (author's interview 2010).

With the push by political entrepreneurs such as Liu, the Shanxi government set environmental protection as a goal that is equally important as economic development. In the "Eleventh Five Year Plan (2006–2010) for Economic and Social Development in Shanxi Province," 44 indicators were set in order to coordinate economic growth and environmental protection, encompassing areas including economic growth, social development, technological advancement, resources and environment, and living standards. Among these indicators, 11 were for economic growth that takes 25 percent in evaluation and nine for environmental protection that takes 20.45 percent.

The economic indicators include the following targets:

1. Ten percent or more increase of annual GDP. By 2010, GDP per capita reaches the national average or higher, with average annual growth rate of 9.6 percent, reduces the gap with national average.
2. The average annual growth of investment in fixed assets increase by 20 percent; total retail sales of consumer goods by 13 percent; international trade by 12 percent; direct investment by 20 percent, with overseas direct investment by 25 percent; annual growth of industrial value added by 14 percent; the proportion of the tertiary industry value added in GDP more than 40 percent; and total provincial fiscal revenue and budgetary revenue growth rate by 15 percent.
3. Inflation growth rate below 2–3 percent.
4. Annual grain outputs over 9 billion kilograms.

The environmental indicators include the following targets:

1. By 2010, 60 million acres of cultivated lands; the rate of forest coverage at 18 percent; green areas in urban residential areas increase by 1 percent annually and reach 35 percent at the end of "11th five-year plan;" the rate of urban household garbage treatment over 60 percent; the rate of urban sewage treatment over 70 percent.
2. In areas with GDP per capita over 10,000 *yuan*, pollutant emissions decrease by 40 percent, with an average annual decrease by 10 percent; the coverage rate of urban central heating system over 80 percent; the rate of coal recycling in mining areas over 60 percent; annual overall energy consumption rate decrease by 5.6 percent, with 25 percent total decrease at the end of the period; and water consumption rate decrease by 8.3 percent, with 35 percent total decrease by the end of the period.

These indicators defined the targets with specific data, thus becoming "hard constraints" for local governments and enterprises to follow.

At the same time, the government emphasized the interaction of industrial planning and environmental enforcement in its policy design. In 2009, Shanxi government published a plan of industrial restructuring and revitalization for eight industries. In 2010 the government launched plan to update equipment manufacturing, modern logistics, energy conservation and environment protection, new energy and new materials, high technology, cultural tourism, and other emerging industries. In planning the traditional pillar industries such as coal, metallurgy, and electric power, each step of the plan such as agenda setting, targets, policy measures, and plan implementation had sufficient portion that stressed environmental protection, energy conservation, and emission reduction. These measures combined the industrial restructuring and revitalization with energy conservation and emission reduction and environmental protection in a holistic way in order to ensure environmental protection is achieved in the process of industrial restructuring.

In order to achieve the environmental protection objectives, Shanxi government tightened environmental protection laws and regulations. Many high energy-consuming and high pollutant-emitting industries and firms were fined, shut down, or restricted in production. These measures produced strong pressures on local governments and firms to reduce energy consumption and emission and upgrade technology.

Shanxi government also adjusted its system of evaluation, using rewards and punishments to hold local governments and firms accountable in implementing the plan. It sets a series of goals to guide the local governments and firms toward the direction of balancing environmental protection and economic development.

5. Assessing the Compatible Development Model

5.1. Improvement of Environmental Protection

Shanxi achieved a balanced development through the adjusted incentive mechanisms. Not only was the amount of pollutant emission reduced and environmental quality improved but, more importantly, economic growth did not slow down as a result of tightened environmental regulations. The progress is reflected in the following indictors.

First, the total investment in environmental protection had grown from 3.2 billion *yuan* in 2003 to 15.78 billion in 2009. Shanxi also purchased 20,205 wastewater treatment facilities and 56,714 waste gas treatment facilities during this period.

Second, the government launched a series of campaigns in order to strengthen environmental enforcement. During these campaigns, the government closed down more than one million polluted firms and facilities and investigated and prosecuted more than 13,000 violation cases.

Third, as a result, Shanxi's pollutant emission amount decreased significantly. COD emissions of industrial wastewater decreased from 16 tons in 2003 to 14 tons in 2008. Industrial SO_2 emissions decreased from 1.18 million tons in 2006 to 1.06 million tons in 2008. Industrial solid waste decreased from 6.29 million tons in 2003 to 2.32 million tons in 2008.

Fourth, the quality of ambient water, surface water, and groundwater has improved significantly. Take Taiyuan city as an example. The PM2.5, SO_2, and NO_2 levels were 0.172, 0.099, and 0.031 in 2003, respectively. By 2008, the figures dropped to 0.106, 0.075, and 0.022.

Among the 11 major cities in Shanxi, Taiyuan was the only city whose air quality did not meet Grade II National Air Quality Standard (which is the second best among six grades). All other major cities have achieved quite impressive results. For example, the number of days with air quality meeting

Grade II Standard from 2008 to 2009 increased by 43 (11 percent) and 22 (6.9 percent) in Datong and Shuozhou, respectively.

5.2. Continued Economic Growth

At the same time that Shangxi enjoyed a better air and water quality thanks to the tougher environmental protection standards and enforcement, its growth, to some theorists' surprise, did not slow down but remained robust. This is particularly impressive given the 2008 global financial crisis that affected the nation. From 2005 to 2009, total GDP, investment in fixed assets, disposable income per capita among urban residents, and net income per capita among rural residents increased by 74, 172, 57, and 47 percent, respectively.

Meanwhile, the economic structure improved when the resource allocation across the primary, secondary, and tertiary industries became friendlier to environment. The growth rate of investment in the primary and the tertiary industries rose considerably, well above 30 percent, while the growth rate of investment in the secondary industry decreased over years, with 15 percent in 2009, compared to 96.6 and 60.7 percent in primary and tertiary industries, respectively.

The last indicator of the structural improvement is the decreased rate of energy consumption as the share of GDP that occurred while the economy continued to grow. The energy consumption per unit of GDP, energy Consumption per unit of industrial enterprises above designated size, and electric power consumption for per unit of GDP dropped by 7.4, 9.3, and 10 percent, from 2007 to 2008, respectively.

6. Conclusion

Shanxi Province was once the most polluted region in China. Today, it becomes one of the regions that have made most remarkable progress in both environmental protection and robust economic growth simultaneously. Shanxi has made this progress in less than five years, which is remarkable in both the Chinese and international standards. The experiences of environmental governance in Shanxi have significant implications for China to explore a better approach to achieve a harmonious development.

Three lessons can be learned from Shanxi's compatible development model. First, Shanxi government strategically utilized its geographical location, resource endowments, economic structure, and environmental capacity to plan and coordinate its economic growth and environmental protection, setting goals and action plans for a holistic development strategy. Second, Shanxi government strategically synergized industry restructuring and environmental protection and implemented its plans in eight industries. At the same time, it strengthened the enforcement for environmental protection while promoting industrial upgrading in order to improve the competitiveness of its firms. Third, a compatible development strategy was reinforced by an improved incentive structure with better assessment and accountability systems with clear rewards and punishments. The combination of economic growth and environmental protection in the assessment and accountability system induced local governments to take a balanced approach in dealing with growth and environment simultaneously.

Although Shanxi's success in environmental governance may be rare in today's China, the problems it faced, however, are not uncommon in most other regions. The experiences of Shanxi therefore have important implications for the entire country. Three suggestions can be made. First, it is critical to create an incentive structure designed for a compatible development model. The success of environmental governance in Shanxi Province was made without the change of economic and political structures. Within the existing institutional framework, the government modified the incentive structures and made environmental protection the responsibility of local governments at each level through the target responsibility system, making local protectionism difficult to intervene. Through the performance evaluation system and reward and punishment system, the government was able to effectively motivate local officials to implement policies.

Second, intergovernmental coordination is critical for enforcement. The effects of environmental pollution cut across several governmental departments. It is difficult for one department to comprehend the consequences of certain type of pollution, which may appear after certain amount of time has passed. However, the consequence of certain type of pollution can be inferred from other indicators such as energy consumption and investment, which can be detected by the respective departments more easily. An important measure in Shanxi's experience was to coordinate different departments and create a mechanism through which different departments share information in order to better understand the consequences, predict the risks of certain types of pollution, and cooperate in enforcement in order to increase efficiency and effectiveness.

Third, better environmental governance must be centered on incentive structures in making policies and finding solutions. Environmental problems are not simply the problems of technology, political power, or individuals. Institutions matter, as the literature of institutionalism has suggested. Environmental governance calls for institutional designs that offer incentives to individuals so they behave in an environment-friendly manner and also make it costly for individuals to pollute or ignore pollution. In other words, a good institution design makes environmental protection an individual choice—a choice an individual finds to be in line with his or her personal interests.

The success of the compatible development model in Shanxi is suggestive. It demonstrates that environmental protection and economic growth do not contradict. Economic growth does not necessarily come at the expense of pollution. Neither does protecting environment necessarily sacrifice economic growth and economic competitiveness, as long as properly designed institutions are in place to make the incentive structures available for this compatible development model.

BIBLIOGRAPHY

Beckerman, Wilfred. 1992. "Economic Growth and the Environment: Whose Growth? Whose Environment?" *World Development* 20(1) (April): 481–496.

Cao, Xin. 2004. "Study on the Relationship of Economic Development and Environmental Protection [*jingjifazhan yu huanjingbaohu guanxi yanjiu*]." *Social Science Journal [shehui kexue jikan]* 2: 60–64.

Georgescu-Roegen, Nicholas. 1971. *The Entropy Law and the Economic Process.* Cambridge: Harvard University Press.

Grossman, Gene and Alan Krueger.1995. "Economic Growth and the Environment." *The Quarterly Journal of Economics* 110(2): 353–377.

Holdren, John and Paul Ehrlich. 1971. "Overpopulation and the Potential for Ecocide." In Holdren and Ehrlich (eds.) *Global Ecology*. New York: Harcourt Brace Jovanovich, pp. 64–78.

Jansson, Annmari, Hammer Monica, Folke Carl and Costanza Robert. (eds.). 1994. *Investing in Natural Capital—The Ecological Economics Approach to Sustainability.* Washington, D.C.: Island Press.

Klevorick, Alvin K. 1996. "The Race to the Bottom in a Federal System: Lessons from the World of Trade Policy." *Yale Law and Policy Review* 14(2): 177–186.

Meadows, Donella, Dennis Meadows, Jorgen Randers and William Behrens III. 1972. *Limits to Growth.* New York: New American Library.

Pi, Jiancai. 2010. "Environmental Protection and Economic Development in the Chinese Decentralized System [*zhongguoshi fenquanxia de huanjingbaohu yu jingjifazhan*]." *Research on Financial and Economic Issues* [*caijing wenti yanjiu*] 6: 10–14.

Porter, Michael E. 1995. "Toward a New Conception of the Environment-Competitiveness Relationship." *Journal of Economic Perspectives* 9(4): 97–118.

Qi, Ye and Lingyun Zhang. 2007. "The Applicability of 'Green GDP' in Officials' Assessment ['luse GDP' zai ganbukaohe zhong de shiyongxing fenxi]." *Chinese Public Administration [zhongguo xingzhengguanli]* 12: 26–30.

Schumpeter, Joseph. 1934. *The Theory of Economic Development: An Inquiry into Profits, Capital, Credit, Interest, and the Business Cycle.* Cambridge: Harvard University Press.

Shanxi Department of Environmental Protection. 2009. "Blue Sky Clean Water Project in Three Years [bishuilantian sannian jian]" *Environmental Protection*.[huanjing biaohu] 11: 1–25.

Shanxi Province—Department of Environmental Protection. 2010. Communique of Shanxi on the 2009 Environmental Quality. http://www.sxhb.gov.cn/news.do?action=info&id=24769

Shanxi Province—Shanxi Statistical Information Website. 2010. Statistical Communiqués of Shanxi Province, 2005 to 2009. Shanxi Economic and Social Development. http://www.stats-sx.gov.cn/html/2010-3/2010310163817172 385702.html

Sun, Changxue. 2006. "Governmental Action and Sustainable Development of Resource and Environment [*zhengfuzuowei yu ziyuanhuanjing kechixu fazhan*]." *Reform of Economic System* [*jingji tizhi gaige*] 1: 31–34.

Wang, Yuming. 2008. "The Practice and Improvement of Leaders Environmental Protection Evaluation System [*guangdong lingdaoganbu huanjingbaohu shiji kaohezhidu de shijianyujiqi wanshan*]." *Journal of Guangdong Institute of Public Administration* [*guangdong xingzhengxueyuan xuebao*]. 2: 21–25.

Xing, Xiufeng and Yingyu Liu. 2006. "Measurement and Analysis of the Relationship between Economic Development and Environmental Protection of Shandong Province [*shandongsheng jingjifazhan yu huanjingbaohu guanxi de jiliangfenxi*]." *China Population Resources and Environment* [*zhongguo renkou ziyuan yu huanjing*] 1: 58–61.

Xiong, Peng.2005."Environmental Protection and Economic Development: The Controversy between Porter Hypothesis and Traditional Neo Classical Economics [*huanjingbaohu yu jingjifazhan-ping botejiashuo yu chuantongxingudian jingjixue zhizheng*]." *Contemporary Economic Management* [*dangdai jingjiguanli*] 5: 80–84.

Zhang, Hongfeng and Feng Zhou. 2009. "Research of Regulation Performance of the Mutual Empowerment of Environmental Protection and Economic Development [*huanjingbaohu yu jingjifazhan shuangying de guizhijixiao shizhengfenxi*]." *Economic Research* [*jingji yanjiu*] 3: 14–26.

Zhang, Lingyun and Ye Qi. 2010. "Explaining Local Environmental Regulation Dilemma: Hypothesis of Political Incentives and Financial Restraints [*difang huanjingjianguan kunjing jieshi-zhengzhijili yu caizhengyueshu jiashuo*]." *Chinese Public Administration* [*zhongguo xingzhengguanli*] 3: 93–97.

Zhang, Man. 2001. "A Strategy for Economic Growth and Environmental protection to Coexist [*jingji fanzh yu huanjingbaohu de gongsheng guanxi*]." *Research on Financial and Economic Issues* [*caijing wenti yanjiu*] 5: 74–80.

Zhou, Jingkun. 2009. "Performance Management System of Party and Government Leaders in Environment Protection Based on the Local Government Environmental Protection Strategy Objectives [*jiyu difangzhengfu huanjingbaohu zhanlue guihua de dangzhenglingdao huanbaojixiao guanlitixi yanjiu*]." PhD diss. Donghua University.

Part II

Firms and Environmental Regulation

CHAPTER 5

ENVIRONMENTAL MANAGEMENT,
FINANCING, AND PERFORMANCE IN
CHINESE FIRMS: EVIDENCE FROM
A NATIONWIDE SURVEY

Xiaojun Li

1. INTRODUCTION

As China's economy and population continue to grow, technological innovations that mitigate the impact of economic development and repair the damage caused by these changes will be critical to China's efforts toward more sustainable development. In response to its environmental problems, China's 12th Five Year Plan includes steps to boost the environmental protection industry by creating targets for environmental standards and financing as well as regulating companies producing core environmental products.

Achieving these environmental goals will depend to a great extent on China's fast-growing industrial sector (Economy 2004), which accounts for 44.7 percent of waste water, 38 percent of carbon dioxide, 79.4 percent of soot, and 74.6 percent of nitrogen oxide released in the country according to the 2006 National Environment Report (Ministry of Environmental Protection 2006). How do Chinese firms perform environmentally? Are firms of certain size and ownership structure greener than others? Is there any regional variation? Answers to questions like these will help the government identify poorly performing firms that are eroding China's effort toward environmental sustainability and formulate targeted and effective policies to improve environmental performance of Chinese firms as a whole.

The primary focus of this chapter, therefore, is to provide an overview of the environmental performance of China's industrial firms, from management and certification to financing and emission control, using data from a recent nationwide survey. The remainder of the chapter is organized in three sections. The second section introduces the survey and the sample composition of the firms. The third section examines the data from the survey in more detail, focusing on the relationship between a battery of firm characteristics and various aspects of environmental performance in Chinese firms. The final section summarizes the major findings, points out the limitation of the survey data, and provides some policy recommendations.

2. Survey and Data

The data used for this chapter have been drawn from a survey about corporate social responsibilities in Chinese firms, conducted jointly by the China Center for Economic Research (CCER) and the National Bureau of Statistics (NBS) in the spring of 2006 (hereafter referred to as the CCER Survey). The survey was administered to 1,268 firms in 12 Chinese cities. Trained enumerators from the local statistics bureau carried out the survey in each city. The survey contains a number of sections. The manager or the person in charge of the environmental department of each firm was asked to answer a number of questions for the section on environmental management in the survey. Their responses were randomly checked with records in the local environmental bureaus for quality control.

The cities in the survey were chosen based on the principle of representation. Overall, the 12 sample cities constitute a reasonable representation of China in terms of geographic locations as well as socio-economic conditions.[1] Beijing and Chongqing are two provincial-level cities. Changchun, Shijiazhuang, Xi'an, and Hangzhou are provincial capitals of Jilin, Hebei, Shannxi, and Zhejiang provinces, respectively. Wujiang and Shunde are county-level cities. The other cities are medium-sized prefecture-level cities. Beijing, Wujiang, Hangzhou, and Shunde belong to the coastal region; Chifeng, Xi'an, Shiyan, and Chongqing belong to the western region; and the rest belong to the central region.

In each city, 100 firms are sampled among those with an annual sales volume larger than five million RMB using a two-stage stratified sampling strategy.[2] The first stratum is firm ownership: state and collectively owned firms, private firms, and foreign firms (including joint ventures). Stated-owned enterprises (SOEs) are those in which the state holds the controlling shares. Private firms include companies with majority private shares. Foreign firms are those where the majority of the shares are held by foreign entities including Hong Kong, Macao, and Taiwanese businesses. The second stratum is firm size according to employment: firms hiring less than 500 people are defined as small firms; those hiring 500–2,000 people are defined as medium firms; and large firms are those that employ more than 2,000 people.[3] This categorization of firm size follows the routines used by the NBS, which is defined by the State Economic and Trade Commission (SETC) in 2003.

Table 5.1 shows the distribution of firms by survey city and a number of firm characteristics. Focusing first on the sampled firms as a whole (the last row of Table 5.1), we see that the majority of them are small in size. Medium and large firms each only account for 14.9 and 12.3 percent of the sample, respectively. With respect to sectors, 63.6 percent of the firms are in pollution-intensive industries: mining, chemical, machinery, and fuel production. In terms of ownership structure, the majority (68.7 percent) of the sampled firms are privately owned. One-fifth of the firms are foreign enterprises and joint ventures. SOEs account for a little over 10 percent of the total sample, reflecting the fact that SOE restructuring has been generally completed by the time of the survey (Oi 2010). Finally, 41.4 percent of the sampled firms are export-oriented. The sample averages of these economic indicators are comparable to numbers reported in the Statistical Report of National Economy and Social Development published by the NBS, suggesting that the sample firms are a relatively good representation of the national average.

There are, however, substantial regional variations. Take ownership structure for an example. The percentage of foreign invested firms is the highest in the coastal region, second in the central region, and the lowest in the western region. Among the four coastal cities, Hangzhou has a heavier presence of private firms, Shunde a heavier presence of foreign invested firms. This latter feature has a lot to do with Shunde's geographic proximity to Hong Kong and Macao. Xi'an is an exception in the western region; its share of foreign firms and joint ventures is comparable to most central cities. However, it also has the highest percentage of SOEs among all the sample cities—over 5 percent more than Beijing, where most of the centrally controlled SOEs are located.

3. Findings

For the purpose of this chapter, I focus on all the questions in the environmental section of the CCER Survey. The response rates of these questions are quite good, ranging from 61.6 to 96.6 percent. The questions in this section can be grouped into four different subsections, each focusing on one aspect of a firm's environmental performance: management, certification, financing, and emission control. In each of the four subsections that follow, I will first present a broad overview of how the surveyed firms perform on the whole. I will then then break the firms down according to region, size, pollution intensity, ownership structure, and export orientation to examine whether certain firms have better environmental performance than others. The main results are presented in Table 5.2.

Table 5.1 Distribution of firms by size, pollution intensity, ownership structure and export orientation in sample cities

Region	City	Firms sampled	Firm size			Pollution intensity		Ownership structure			Export firm	
			Small (%)	Medium (%)	Large (%)	High (%)	Low (%)	State (%)	Private (%)	Foreign (%)	No (%)	Yes (%)
Coastal	Beijing	100	80.8	14.1	5.1	67.7	32.3	17.2	43.4	39.4	73.7	26.3
	Hangzhou	100	65.0	24.0	11.0	78.0	22.0	1.1	73.7	25.2	35.0	65.0
	Wujiang	100	58.0	17.0	25.0	61.0	39.0	2.6	52.6	44.9	30.0	70.0
	Shunde	115	63.5	21.7	14.8	64.4	35.6	0.9	57.8	41.2	30.4	69.6
Central	Changchun	100	75.0	5.0	20.0	46.0	54.0	10.3	81.6	8.0	70.0	30.0
	Shijiazhuang	100	77.0	11.0	12.0	68.0	32.0	12.0	77.2	10.8	58.0	42.0
	Dandong	109	83.5	11.0	5.5	65.1	34.9	14.0	67.3	18.7	50.5	49.5
	Zibo	140	82.9	8.8	8.6	83.6	16.4	8.8	65.4	25.7	59.3	40.7
Western	Chongqing	100	85.0	13.0	2.0	50.0	50.0	10.0	84.0	6.0	81.0	19.0
	Xi'an	101	50.5	33.7	15.8	57.4	42.6	22.7	58.8	18.6	56.4	43.6
	Chifeng	100	81.8	7.1	11.1	74.7	25.3	11.1	88.9	0.0	79.8	20.2
	Shiyan	115	67.0	14.6	18.5	39.8	60.2	15.6	77.8	6.7	83.5	16.5
Total		1268	72.8	14.9	12.3	63.6	36.4	10.5	68.7	20.8	58.6	41.4

Note: Firm size is determined by the number of employment; pollution-intensive industries include mining, chemical, machinery, and fuel production; ownership is determined by the controlling share in the firm.
Source: CCER Survey 2006.

Table 5.2 Environmental management, certification, financing and performance in Chinese firms

		Environmental management				Environmental certification		Environmental financing		Environmental performance			
		(1)	(2)	(3)	(4)	(5)	(6)	(7)	(8)	(9)	(10)	(11)	(12)
Region	Coastal	52.12	6.52	25.15	2.11	21.15	13.84	182.10	40.25	79.02	81.37	79.45	77.96
	Central	49.42	6.67	26.11	2.93	10.61	12.84	122.71	58.20	78.32	80.12	77.42	79.06
	Western	58.82	11.13	36.31	2.39	17.24	10.70	171.64	29.61	81.52	82.19	83.96	75.24
	Test Stat	7.63**	3.12**	12.38***	0.22	16.02***	1.68	0.76	0.99	1.00	0.45	4.22	1.41
Firm size	Small	47.32	4.77	25.56	1.63	10.47	12.13	56.20	20.35	77.10	78.75	77.58	75.86
	Medium	76.11	13.43	37.74	3.31	29.31	11.24	284.99	48.62	84.38	85.88	85.43	81.82
	Large	61.74	19.85	39.06	6.63	33.33	16.30	595.69	162.19	85.37	87.40	87.10	80.62
	Test Stat	54.76***	21.21***	16.21***	5.12***	70.09***	2.14	40.53***	15.52***	7.08**	8.21**	9.00**	3.40
Pollution intensity	High	54.88	7.59	69.26	2.92	13.94	11.57	150.59	42.36	77.68	78.54	78.11	75.91
	Low	50.56	8.84	73.39	1.77	20.00	14.08	171.40	44.37	83.01	85.71	83.78	80.11
	Test Stat	2.13	0.50	1.93	1.15	6.96***	1.54	0.22	0.01	3.54*	7.45***	4.23**	2.24
Ownership	State	63.33	13.5	73.00	2.29	20.00	11.82	329.02	96.35	65.26	67.35	62.35	64.29
	Private	50.58	6.34	70.86	2.68	12.92	12.04	119.35	37.50	79.07	80.64	79.78	76.23
	Foreign	58.23	11.18	69.39	1.94	25.45	14.35	223.95	38.15	89.08	90.10	90.40	89.07
	Test Stat	9.75***	4.90**	0.42	0.17	21.45***	0.92	5.12***	2.01	21.72***	22.40***	28.73***	24.46***
Export	Yes	57.43	10.39	68.46	3.55	22.10	15.02	214.93	58.73	83.47	85.01	83.55	82.19
	No	50.48	6.38	72.39	1.80	11.98	10.78	119.65	32.28	76.70	78.31	77.80	74.15
	Test Stat	5.73**	5.42**	1.81	2.78**	20.29***	4.61**	4.91**	2.33	6.11**	6.84***	4.55**	8.54**
Total		53.31	8.05	70.79	2.5	16.16	12.48	158.12	43.10	79.55	81.18	80.22	77.46

Note: The columns for environmental management are (1) designated environmental protection or management department/office; (2) number of staff working in positions related to environmental protection; (3) compilation of annual environmental performance or sustainability report; and (4) number of environmental management training sessions in 2005. The cells in each of the two columns for environmental certifications denote the percentage of firms that are (5) certified by ISO14000 and (6) national environmental labels. The two columns for environmental financing are the amount of money (in 10,000 RMB) a firm spent on (7) pollution abatement technologies, including equipment purchase, maintenance, and technology innovation for emission reduction and environmental protection and (8) environmental operational costs (e.g., audit, supervision, fines, etc.) from 2003 to 2005. The cells in each of the four columns for environmental performance denote the percentage of firms that meet national emission standards in (9) gas, (10) water, (11) solid waste, and (12) noise. Test statistics of (1), (3), (5), (6), (9), (10), (11), and (12) are based on Chi-square tests. Test statistics for (2), (4), (7), and (8) are based on ANOVA/two-sample *t*-tests. * $p < 0.1$, ** $p < 0.05$, *** $p < 0.01$.
Source: CCER Survey 2006.

3.1. Environmental Management

The following four questions tap into environmental management in Chinese firms:

1. Does the firm have an office/department designated for environmental protection or management?
2. How many people in the firm hold positions related to environmental protection?
3. Does the firm compile an annual environmental performance or sustainability report?
4. In 2005, how many training sessions for environmental management did the firm organize or participate in?

Overall, a little more than half (53.3 percent) of the firms have designated environmental protection or management departments, with an average staff of eight people, or 1.07 percent of the total employment. Figure 5.1 plots the year when the 626 firms set up their environmental protection or management departments. Between 1949 and 1978, the year when the reform and opening policy officially rolled out, only 35 firms established offices or departments for environmental protection and management. In contrast, 384 firms set up their environmental departments in the 2000s. Part of the reason that few firms have designated environmental offices in the early years may be attributed to the fact that many of the firms did not exist back then.[4] Still, the sharp increase in designated environmental protection and management departments can be attributed to the firms' response to the state's goal to improve environmental sustainability in the industrial sector in the new millennium.

A quarter (25.22 percent) of the firms have compiled annual environmental performance or sustainable production reports. The majority of these firms, however, did not do so until after 2000 (see Figure 5.2), which is similar to the pattern in the establishment of environmental protection departments described above. Interestingly, 159, or 13.68 percent, of the firms say that they have never heard of the annual environmental performance or sustainable production report. Finally, the surveyed firms on average have held or participated in 2.5 environmental training sessions in 2005.

Is there any variation in environmental management across firms? Columns (1)–(4) of Table 5.2 disaggregate the responses to these four questions by region, size, pollution intensity, ownership structure, and export orientation. First, more firms in the western region have designated environmental protection and management departments and compile annual environmental protection reports, compared to those in the coastal and central regions. Similarly, western firms, on average, also have more people working in environment-related positions. These differences are statistically significant. On the other hand, there is no discernible difference in the number of training sessions held by firms in these three regions.

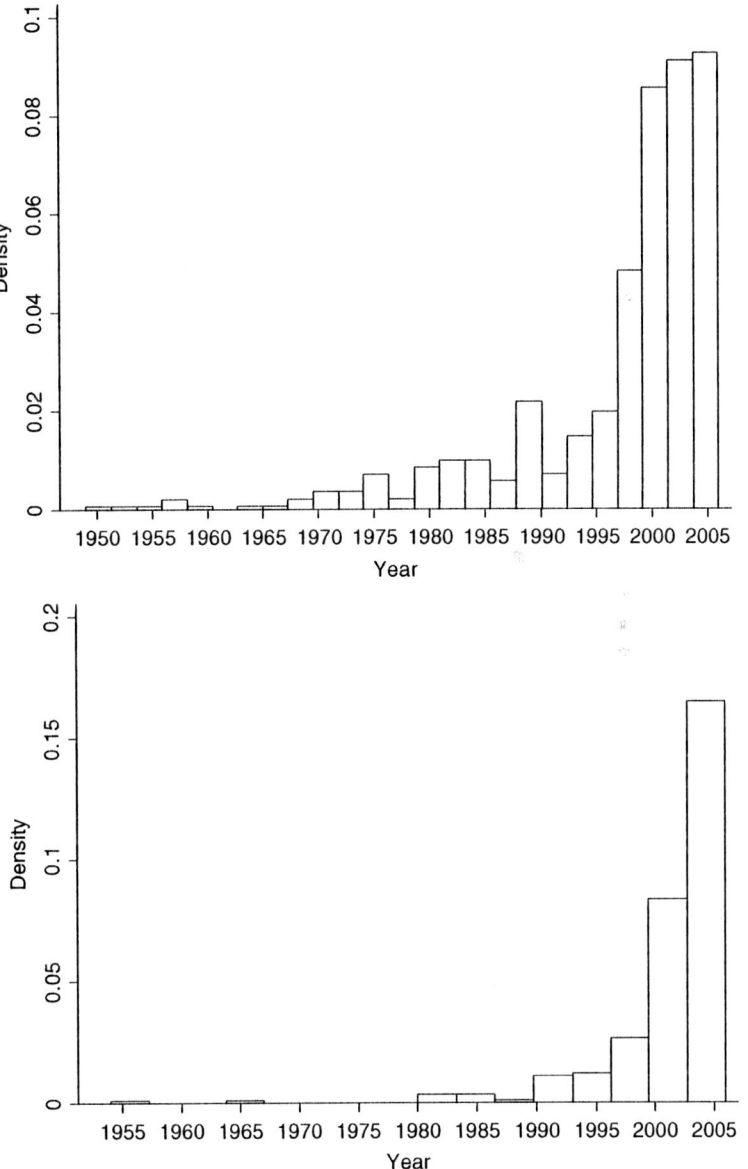

Figure 5.1 Environmental departments and annual environmental reports by year of establishment and first publication
Source: CCER Survey 2006.

Second, firm size matters a lot. Large firms perform better in three of the four measures (staff, report, and training), while designated environmental offices are most likely found in medium-sized firms.[5] Third, there is no relationship between a firm's pollution intensity and its performance

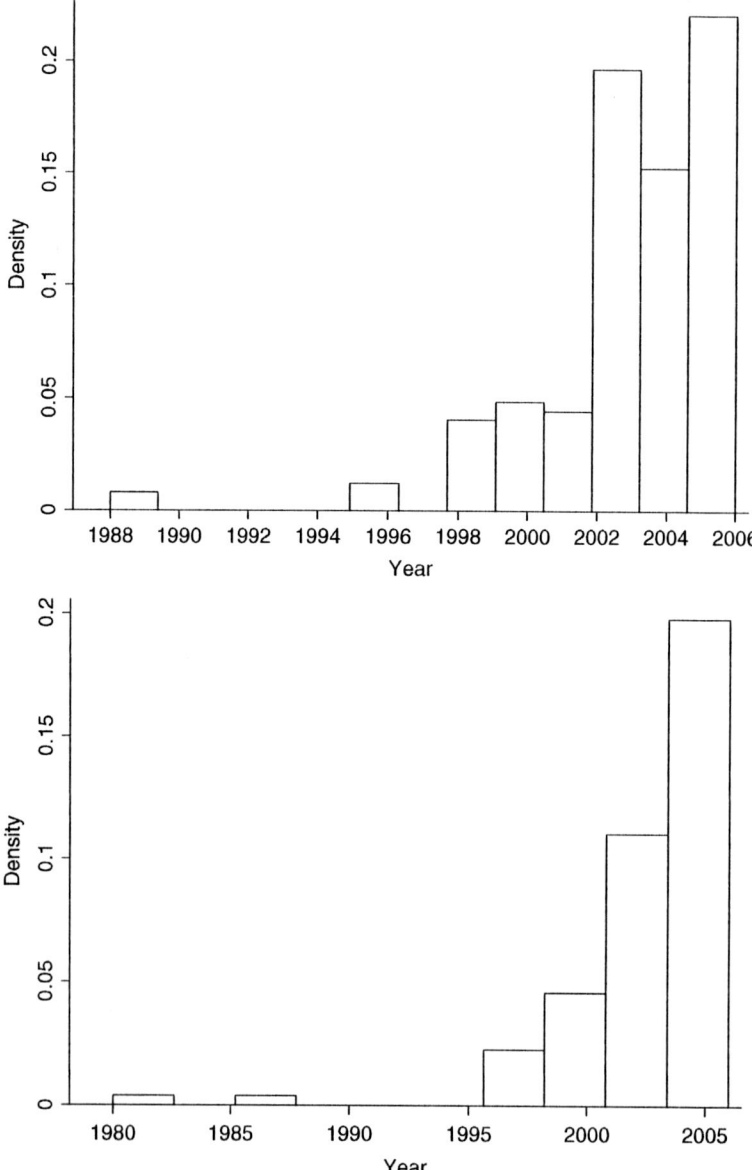

Figure 5.2 ISO14000 and national environmental labeling by year of certification
Note: The ISO14000 program was established in 1996 and the National Environmental Labeling program was established in 2003.
Source: CCER Survey 2006.

on environmental management. Finally, ownership structure and export orientation matter only for environmental office and staff. SOEs are more likely to have established offices or departments designated for environmental protection and management and employ more people in environmental management positions compared to private firms and joint ventures. Same can be said of export firms.

3.1.1. Certification

I examine two environmental certification programs in Chinese firms using the following two questions:

1. Is the firm certified by ISO 14,000?
2. Has the firm obtained national environmental labeling for its products?

Established by the British Standards Institution in 1996, ISO14000 is a core set of standards used by organizations for designing and implementing an effective environmental management system. The China Environmental Labeling Program was launched by the China Environmental United Certification Center (formerly the Environmental Protection Administration Environmental Certification Center) in 2003. These two certification programs thus represent both international and domestic endorsement that the certified firms and their products are managed and manufactured with high environmental standards and quality.

Overall, 16.1 percent of the firms are ISO14000 certified and 12.5 percent of the firms have obtained national environmental labeling for their products. Figure 5.2 shows the number of certifications over time. For the 180 firms with ISO14000, the majority (91.7 percent) were certified in the 2000s. Similarly, 89 of the 101 firms with national environmental labels obtained their labels in the 2000s. Ironically, a couple of firms say that they have been certified by ISO14000 or the national environmental labeling program in the 1980s, well before those programs were officially established.

Columns (5) and (6) of Table 5.2 once again disaggregate the responses to these two questions by region, size, pollution intensity, ownership structure, and export orientation. For ISO14000 certification, the differences among firms in all five categories are statistically significant. Firms in the coastal region fare the best (21.2 percent), followed by those in the western (17.2 percent) and central (10.6 percent) regions. One-third of the large firms are certified, which is closely followed by medium-sized firms (29.3 percent). Small firms, however, lag far behind, with a little more than 10 percent of them certified by ISO14000. In terms of ownership, only 12.9 percent of the private firms are ISO14000 certified while 25.5 percent of the foreign firms are certified. State firms hover between the two, with 20 percent of the SOEs certified by ISO14000. Finally, pollution-intensive (20 percent) and export firms (22.1 percent) are both more likely to be certified.

There is very little variation when it comes to national environmental labeling. Coastal, large, and foreign firms are slightly more likely to have obtained national environmental labeling than other firms, but the differences are not statistically significant. The exceptions are export firms, which are significantly more likely to have obtained national environmental labels for their products. While this may seem counterintuitive at first, it should be pointed out that many export firms also sell their product in the domestic market, so getting the national environmental labels are useful for them as well. In addition, export firms are more likely to be certified by ISO14000, which presumably also makes it easier for them to obtain the national standards.

3.1.2. Financing

I use two questions to look at environmental financing. The first question asks how much the firm spent in the past three years (2003–2005) on pollution abatement technologies, including equipment purchase, maintenance, and technology innovation for emission reduction and environmental protection. On average, firms spent 1.58 million RMB on pollution abatement technologies and equipment from 2003 to 2005 (see columns (7) and (8) of Table 5.2). More specifically, firms in the coastal region invest the most (1.82 million RMB) followed by those in the western (1.72 million RMB) and central (1.23 million RMB) regions. SOEs invest the most (3.29 million RMB) followed by foreign firms (2.24 million RMB) and private firms (1.19 million RMB). Large firms invest the most (5.95 million RMB) on pollution abatement technologies and equipment, twice and 11.5 times as much as the amount spent by medium and small firms, respectively. Export-oriented firms spend almost twice as much on pollution abatement technologies and equipment as import-competing firms. Finally, pollution intensity leads to more spending, but the difference is not statistically significant.

The second question asks about the firm's environmental operational costs such as audit, supervision, fines, and other administrative fees from 2003 to 2005. The average operational cost is 0.43 million RMB. The degree and direction of variation in operational costs among firms of different size, pollution intensity, ownership structure, and export orientation are similar to the variation in investment, though only the difference in size is statistically significant. In contrast, the relationship between region and operational costs are almost the opposite. Firms in the central region incur the most operational costs (0.58 million RMB), followed by those in the coastal (0.4 million RMB) and western (0.3 million RMB) regions. Nevertheless, the differences are not statistically significant.

3.1.3. Environmental Standards and Emission Controls

The final set of questions look at the environmental performance of Chinese industrial firms. Specifically, the survey asks whether or not the firm meets the national emission standards in gas (including CO_2 and SO_2), waste water, solid waste, and noise. On average, over four quarters of the surveyed firms say that they are able to meet these standards (see columns (9)–(12) of Table 5.2). Among these firms, 7.68 percent feel the standards are too strict and 0.4 percent too lenient. The rest say that the national standards are about right. There are 5.2 percent of the firms that say they encounter a lot of difficulties in meeting the national emission standards, 46.7 percent of the firms indicate some difficulties, and the rest indicate no difficulty at all. Of the firms, 53.9 percent say that they invested additional resources in order to meet these standards.

There is considerable variation in the ability to meet national standards with respect to firm size, pollution intensity, ownership structure, and export orientation. Large firms perform better than medium and small firms in meeting national emission standards of gas, waste water, and solid waste. Not surprisingly, pollution-intensive firms are less likely to meet national emission standards in three of the four categories. In terms of ownership structure, foreign firms are the most likely to meet all four national emission standards, followed by private firms and then SOEs. Finally, export-oriented firms are more likely to meet all four national emission standards compared to import-competing firms.

4. Discussions

This chapter uses a recent nationwide survey in China to examine the environmental performance of industrial firms. Overall, the results show that Chinese firms at the time of the survey were well aware of national environmental regulations and standards and have invested a sizable amount of financial and human resources on environmental management and pollution abatement technologies. The majority of the firms are able to meet national emission standards, but only a small percentage of the firms have been certified by national and/or international environmental standards such as the ISO14000.

There is a lot of variation among the firms, however, with respect to geographical location, size, pollution intensity, ownership structure, and export orientation. In particular, large firms consistently perform better than medium and small firms. They invest more on environmental protection, are better managed, and are more likely to meet national environmental and emission standards. This is probably because these large, publicly visible firms are more likely to be carefully scrutinized by environmental authorities. At the same time, they are also more likely to benefit from economies of scale in pollution abatement technologies, leading to better environmental performance overall (Dasgupta et al. 2000). Export-oriented firms also perform better in environmental management, financing, and emission control. This may be explained by the fact that, in developing countries such as China, these firms often engage in self-regulation to prevent the use of environmental regulations in developed countries as protective trade barriers (Rugman and Verbeke 1998).

The most interesting finding is the relationship between ownership structure and environmental performance. SOEs perform the worst environmentally *despite* more resources spent on environmental technologies and management compared to firms of other ownership structure. Why? There are several potential explanations.

First, studies on other countries have found that state ownership is strongly associated with higher levels of pollutant intensity and less investment in pollution abatement technologies because SOEs have a record of wasteful resource use and financial distress and are often shielded from formal and informal regulatory pressure (Dasgupta et al. 2000; Hettige et al.1996; World Bank 1999). The same logic can be applied to China. The SOEs in China possess many of the country's industrial inefficiencies due to "problems in corporate governance, weak management skills and organizational structures, power monopolies, subsidies, and preferential policies" (Serger and Breidne 2007). Under the planned economy, SOEs have historically provided for all its workers, which constitute a large social and financial burden (Oi 2005). Because of these burdens that are the legacy of China's socialist system, these firms are slow to change and do not have as many incentives to be flexible and adopt new, cleaner technologies.

In addition, SOEs have more environmental bargaining power, which is defined as a firm's capacity to negotiate with the local or national environmental agencies pertaining to the enforcement of pollution control regulations (Wang et al. 2003). Enforcement is particularly problematic in China as local environment agents are appointed by local governments whose top priority is economic growth (Jahiel 1998; Tong 2007). Having strong connections and close ties with the growth-oriented local leaders, SOEs are often able to elicit lower pollution payments or punishments—sometimes nothing at all—and thus have less incentive to comply with environmental standards.

Meanwhile, although private firms are driven by economic profitability, often at the expense of environmental sustainability (Earnhart and Lubomir 2010), there are at least two reasons why they are more likely to perform environmentally better than SOEs. First, from the economic perspective, private firms are more efficient in resource utilization and less polluting in the first place. Talukdar and Meisner (2001), for instance, find a significantly negative relationship between the degree of private sector involvement and CO_2 emission levels in 44 developing countries from 1987 to 1995. Similarly, Caporale et al. (2010) show that private plants in Romania make more significant abatement efforts due to increased competitiveness and efficiency.

Furthermore, private firms have less environmental bargaining power compared to the SOEs, and thus face more stringent enforcement from the local and national environmental authorities (Wang et al. 2003; Wang and Wheeler 2003). In China, the pollution charge has been one of the most important instruments in regulating industrial pollution. In 2004, five billion RMB were collected from more than half a million industrial firms, and the fines are increasing each year (Wang and Jin 2007). For private firms, particularly the small ones, therefore, it would be beneficial in the long run to invest more in clean technology and environmentally sound practices now and avoid future costs from environmental fines and penalties.

Finally, the fact that the foreign joint ventures have the best environmental performance provides no evidence for the "pollution havens" hypothesis, i.e., pollution-intensive multinational corporations (MNCs) from developed countries tend to be reallocated to developing countries with lax environmental standards (e.g., Bao et al. 2010; Copeland and Taylor 2004; Xing and Kolstad 2002; Zeng and Eastin 2007). This is in line with most empirical studies to date that find no significant evidence to support the pollution haven hypothesis (e.g., Dasgupta et al. 2000; Dean et al. 2009; Kirkpatrick and Shimamoto 2008).

Even if MNCs exploit pollution havens, there are at least two reasons why they may be motivated to pollute less than domestic firms, and SOEs in particular. On the one hand, foreign joint ventures enjoy relatively large amounts of knowledge-based, intangible assets such as advanced technologies and managerial skills. Because technologies for clean energy and pollution control usually require relatively sophisticated technological inputs, it is logical to expect that foreign firms should be relatively efficient producers and consumers of goods and services that promote energy efficiency and pollution reduction compared to domestic enterprises. For instance, in assessing the extent to which foreign ownership influences the energy intensity of firms in Cote d'Ivoire, Mexico, and Venezuela, Eskeland and Harrison (2003) find that in each case foreign ownership reduces the energy intensity of plants. Evidence from Cole et al. (2006) also suggests that Japanese firms with FDI tend to pollute less and manage emissions better than firms without FDI. On the other hand, the environmental standards in the source countries of foreign direct investment are usually higher than the developing host countries. Through pressure from environment groups in the home country, therefore, multinationals may engage in self-regulating behavior that raises the standards of environmental protection for all subsidiaries (Christmann and Taylor 2001).

These findings need to be qualified due to a number of limitations in the survey data. First, many of the outcome variables are based on firms' own responses and thus suffer from problem of validity common to all surveys relying on self-reports. If firms of certain type (e.g., small or private) systematically exaggerate their environmental spending and/or the extent of compliance to environmental standards, the results may be biased. Second, since I only have access to the data but no control over its design and implementation of the survey, questions that can further elucidate the competing mechanisms behind the variation in environmental performance such as the frequency and amount of fines paid by the firm were not included in the questionnaire. Finally, the survey was conducted in 2006 and is cross-sectional in nature. Thus, we are unable to make causal arguments regarding which factors contribute to the increase in environmental performance. Nor can we generalize the findings to speak about trends in recent years or change over time.

Despite these limitations, the survey allows us to draw a fairly reliable picture of environmental performance in Chinese industrial firms back in 2006 and offers some clues on what needs to be done in the future. For policymakers in China, the most important lesson is to raise the environmental performance of small and state firms. Doing so will be essential in achieving China's goal of environmental sustainability.

Notes

1. For a more detailed discussion that establishes the national representativeness of the surveyed cities, see supplemental materials provided by the principal investigator of the survey. http://www.chinasurveycenter.org/csdn_en/DownLoadChannel_new/detail.aspx?ClassID=4&DataID=8
2. In some cities, more than 100 firms were sampled. See Table 5.1.
3. The average size of the sample firms was 743 workers in 2005. But the variation was large. The smallest only had two people (it might be a firm that was not in operation), and the largest had 86,991 people.
4. Unfortunately, the survey did not ask when the firm was established.
5. The difference between small and large firms remain even after employment size is controlled.

Bibliography

Bao, Qun, Yuanyuan Chen and Liang Song. 2010. "Foreign Direct Investment and Environmental Pollution in China: A Simultaneous Equations Estimation." *Environment and Development Economics* 16: 71–92.

Caporale, Guglielmo Maria, Christophe Rault, Robert Sova and Anamaria Sova. 2010. Determinants of Pollution Abatement and Control Expenditure in Romania: A Multilevel Analysis. CESifo Working Paper No. 3255.

Christmann, Petra and Glen Taylor. 2001 "Globalization and the Environment: Determinants of Firm Self Regulation in China". *Journal of International Business Studies* 32(3): 439–458.

Cole, Matthew A., Robert J. R. Elliott and Kenichi Shimamoto. 2006 "Globalization, Firm-Level Characteristics and Environmental Management: A Study of Japan." *Ecological Economics* 59(2): 312–323.

Copeland, Brian R. and Taylor M. Scott. 2004. "Trade, Growth, and the Environment." *Journal of Economic Literature* 42(1): 7–71.

Dasgupta, Susmita, Hettige Hemamala, Wheeler David. 2000. "What Improves Environmental Performance? Evidence from the Mexican Industry." *Journal of Environmental Economics and Management* 39: 39–66.

Dean, Judith M., Mary E. Lovely and Hua Wang. 2009. "Are Foreign Investors Attracted to Weak Environmental Regulations? Evaluating the Evidence from China". *Journal of Development Economics* 90(1): 1–13.

Earnhart, Dietrich and Lubomir Lizal. 2010. "Effect of Corporate Economic Performance on Firm-Level Environmental Performance in a Transition Economy." *Environment Resource Economics* 46: 303–329.

Economy, Elizabeth. 2004. *The River Runs Black: The Environmental Challenge to China's Future*. Ithaca, NY: Cornell University Press.

Eskeland, Gunnar. A. and Harrison, Ann E. 2003. "Moving to Greener Pastures? Multinationals and the Pollution Haven Hypothesis." *Journal of Development Economics* 70: 1–23.

Hettige, Hemamala, Mainul Huq and David Wheeler. 1996. "Determinants of Pollution Abatement in Developing Countries: Evidence from South and Southeast Asia." *World Development* 24(12): 1891–1904.

Jahiel, Abigail. 1998. "The Organization of Environmental Protection in China." *The China Quarterly* 156: 757–787.

Kirkpatrick, Colin and Kenichi Shimamoto. (2008). "The Effect of Environmental Regulation on the Locational Choice of Japanese Foreign Direct Investment." *Applied Economics* 40(11), 1399–1409.
Ministry of Environment Protection. 2006. *National Environment Report.* http://zls.mep.gov.cn/hjtj/qghjtjgb/200709/t20070924_109497.htm, accessed November 28, 2010.
Oi, Jean. 2005. "Patterns of Corporate Restructuring in China: Political Constraints on Privatization." *The China Journal* 53: 115–136.
Oi, Jean. 2010. "Navigating Political Cross Currents in China's Corporate Restructuring," In Oi, Jean, Scott Rozelle and Xueguang Zhou (eds.) *Growing Pains: Tensions and Opportunity in China's Transformation.* Stanford, CA: Walter H. Shorenstein Asia-Pacific Research Center Rugman, Alan and Alain Verbeke. 1998. "Corporate Strategy and International Environmental Policy." *Journal of International Business Studies* 29(4): 819–833.
Serger, Sylvia S. and Magnus Breidne. 2007. "China's Fifteen-Year Plan for Science and Technology: An Assessment." *Asia Policy* 4: 135–164.
Talukdar, Debabrata and Meisner Criag M. 2001. "Does the Private Sector Help or Hurt the Environment? Evidence form Carbon Dioxide Pollution in Developing Countries." *World Development* 29(5): 827–840.
Tong, Yanqi. 2007. "Bureaucracy Meets the Environment: Elite Perceptions in Six Chinese Cities." *The China Quarterly* 189: 100–121.
Xing, Yuqing and Charles D. Kolstad. 2002. "Do Lax Environmental Regulations Attract Foreign Investment?" *Environmental and Resource Economics* 21(1): 1–22.
Wang, Hua, Mamingi Nlandu, Laplante Benoit and Dasgupta Susmita. 2003. "Incomplete Enforcement of Pollution Regulation: Bargaining Power of Chinese Factories." *Environmental and Resource Economics* 24: 245–262.
Wang, Hua and Wheeler David. 2003. "Equilibrium Pollution and Economic Development in China." *Environment and Development Economics* 8: 451–466.
Wang, Hua and Yanhong Jin. 2007. "Industrial Ownership and Environmental Performance: Evidence from China." *Environmental and Resource Economics* 36: 255–273.
World Bank. 1999. *Greening Industry: New Roles for Communities, Markets and Governments.* New York: Oxford University Press.
Zeng, Ka and Josh Eastin. 2007. "International Economic Integration and Environmental Protection: The Case of China." *International Studies Quarterly* 51: 971–995.

CHAPTER 6

MIND THE GAP: THE ROLE OF
FOREIGN-INVESTED FIRMS IN
NARROWING THE IMPLEMENTATION
GAP IN CHINA'S ENVIRONMENTAL
GOVERNANCE

Phillip Stalley

1. INTRODUCTION

In March, 2013, Xi Jinping and the Chinese Communist Party's fifth generation of leadership officially assumed power. Among the many challenges facing the new administration, few are more significant than pollution control. Although Beijing has rolled out an assortment of well-intentioned environmental protection laws over the last two decades, environmental indicators are nonetheless grim. Whether one looks at figures related to water, air, or soil pollution, the numbers are bleak. As much as 70 percent of China's rivers, lakes, and reservoirs are affected by pollution (Greenpeace 2011:6). About 90 percent of the river sections around urban areas is considered seriously polluted (World Bank 2007:6), and more than half the groundwater monitored in 182 Chinese cities is deemed undrinkable, largely due to the influence of untreated sewage, industrial pollution, and pesticides (Bloomberg 2011). Soil pollution is also extensive—so much so that soil data were recently treated as a "state secret" by the Ministry of Environmental Protection (MEP) (Zhang 2013). One survey estimates that 10 percent of farmland has levels of heavy metal such as lead and zinc that exceed government limits (Buckley 2011); others place the figure closer to 40 percent (Watts 2012).

Air pollution numbers are similar. In early 2013, Beijing's "airpocalypse" garnered a great deal of media attention as air pollution readings shot quite literally off-the-charts and reached a level 25 times that considered safe in the United States (Lim 2013). For the month of January, Beijing's average concentration of PM 2.5, small particulates that become embedded in the lungs and increase the risk of heart and lung disease, was higher than that of a smoking lounge in a US airport (Waldmeir 2013)! One Chinese scientist estimated that PM 2.5 emissions increased 3 percent per year between 2006 and 2010 (Hook 2013). And thick haze is not limited to Beijing, or even to northern China. According to a 2013 report by Renmin University's School of Environment and Natural Resources, almost 90 percent of China's cities have "heavily" or "severely" polluted air (Yu 2013).

This environmental degradation poses an obvious threat to human health. An estimated 12 million tons of grains are contaminated by heavy metals each year (Xu 2007). The World Bank (2007) determined that outdoor air pollution leads to 350,000–400,000 premature deaths per year (with another 400,000 stemming from indoor air pollution). A more recent study placed the figure much higher estimating that in 2010 outdoor air pollution resulted in 1.2 million premature deaths (Wong 2013a). Over the last three decades, the number of deaths related to lung cancer has risen by 465 percent, despite smoking rates holding steady (Huang 2013). Recently, the MEP admitted that the release of harmful chemicals has given rise to "cancer villages," a phenomenon in which pollution is so extreme that there have been sharp spikes in cases of stomach cancer and which Chinese environmentalists have been documenting for several years (BBC 2013). Little wonder, then, that environmental protests have become increasingly common in China's cities. In 2012, there was a series of successful protests in cities such as Ningbo, Shifang, and Qidong, each of which targeted pollution-intensive industrial projects. These demonstrations followed on the heels of similar protests in previous years and they point to the growing threat environmental deterioration poses to social stability.

While there are many causes of pollution, weak implementation of environmental standards is perhaps the most prominent. Simply put, China suffers from a significant compliance gap in industrial environmental regulations, which Lo et al. describe as "the well documented reality that the rather rigorous requirements that are currently on the books are ubiquitously being flaunted" (Lo et al. 2012:228). Focusing on air quality, the 2007 OECD Environmental Performance Review summarizes the dilemma succinctly: "While China has put in place a legislative and regulatory framework to combat air pollution, enforcement of the rules by the local Environmental Protection Bureaus (EPBs) is still far from adequate" (OECD 2007:69). Although the implementation gap is difficult to quantify or measure precisely, it is nonetheless apparent. A 2005 survey of 509 cities showed that only 23 percent of factories treat sewage before disposing of it; approximately one-third of industrial wastewater and two-thirds of household sewage is released untreated (Economy 2007). Ma Zhong, Dean of Renmin University's School of the Environment and Natural Resources, calculated that the amount of water flowing into Chinese industry was four times the reported amount of waste water output. This translates into 16 billion tons of missing waste water, much of which was assumed to be pumped into the ground illegally (Feng 2013).

Estimates about the number of firms in compliance with discharge standards sometimes can be as low as 10 percent (Warwick 2003). Small firms are often not monitored directly. For instance, EPBs produce estimates of the emissions from small power plants based either on the plants' own reports or on fuel data (OECD 2007:69). One inspection of printed circuit board manufacturers in Dongguan, where much of the information technology sector is concentrated, found that 27 of 41 investigated companies violated environmental law (Friends of Nature, Institute of Public and Environmental Affairs, and Green Beagle 2010). The title of a recent report from China's Central Television makes clear the challenge of enforcement: "Close to 70% of Corporations Using Heavy Metals Suspected of Breaking the Law" (Friends of Nature et al. 2011). Looking at the chemical industry, a researcher at Renmin University estimated that there had been approximately 3,600 chemical spills between 1970 and 2010 (Wong 2013b). That's the equivalent of one spill every four days for 40 consecutive years!

And, of course, the lack of enforcement matters greatly as industry is a major source of China's pollution. In the early 2000s, it was projected that rural firms, many of which are small enough to escape regulators' attention, collectively emitted two-thirds of China's air and water pollution (Tilt 2007:916). Today, about 20 percent of the organic pollutants from all sources are accounted for by discharges from industry (Greenpeace 2011:6, 14). Industry also contributes roughly 50 percent of PM 2.5, more than half of both PM 10 and sulfur dioxide, and just over 70 percent of total suspended particulates (TSP) (Greenpeace 2012a:9, Graph 5).

Pollution is not the only thing China possesses in abundance. It also has one of the world's largest stocks of foreign direct investment (FDI). In most years, China is second only to the United States in attracting FDI. Over the last three decades, for example, Hong Kong companies have established approximately 75,000 factories just in the Pearl River Delta Region, employing no fewer than 11 million workers (Yu and Wong 2011:80). In 2011 alone, China attracted a record $116 billion in FDI as more than 27,000 foreign-invested enterprises were approved by government authorities (Wang and Edwards 2012; Xinhua 2012). As of 2010, there were an estimated 445,000 foreign-invested enterprises (FIE) in China employing just less than 16 percent of the urban workforce (more than 55 million workers) and accounting for approximately a quarter of China's industrial output (Morrison 2013:13). According to the Ministry of Commerce, 480 of the world's top 500 companies have established subsidiaries in China (Xinhua 2008). In 2009 China became the world's leading export nation and today foreign-invested firms account for roughly half of all China's exports (Ross 2012). In short, industrial manufacturing is a critical part of the Chinese economy, a major contributor to China's pollution, and an area in which foreign investors are deeply entrenched.

The significant role played by foreign investors in China's manufacturing sector, combined with China's seemingly intractable pollution challenge, raises important questions about the environmental impact of FDI. To what extent does foreign investment contribute to China's pollution challenge? Do foreign investors exploit China's lax regulatory environment and undermine environmental regulations? Stated simply—to what extent are foreign companies a cause of China's pollution problem? These questions tap into a larger scholarly debate about the relationship between economic integration and environmental regulations. Many scholars and environmentalists express a concern that an overreliance on foreign direct investment risks turning developing countries like China into pollution havens. Attracted by weak regulations, foreign investors use their mobility and deep pockets as leverage over jurisdictions thirsty for capital. This clout allows them to skirt China's environmental regulations or, at the very least, minimize their own pollution abatement expenditures.[1] Foreign multinational corporations (MNC) may produce glossy corporate environmental reports, but it is largely window-dressing for the purpose of greenwashing. From this perspective, foreign investors cause exacerbation of the challenge of narrowing the enforcement gap.[2]

Others paint a very different picture in which foreign investment, and economic openness more generally, ameliorate China's pollution dilemma. Zeng and Easton's volume, *Greening China* (2011), makes this argument most extensively. They argue that foreign companies in China face a variety of pressures to adhere to, and in many cases go beyond, China's environmental standards. Foreign-invested firms that produce for export must make sure they do not lose access to overseas markets where environmental regulations tend to be more stringent. Foreign companies seeking to access China's domestic market are driven by activist pressure and image concerns as "the potential gains from establishing positive public relations with consumers outweigh any savings generate by environmentally destructive behavior" (23). The result is that foreign firms typically self-regulate their environmental protection practices and comply, or even go beyond compliance, with environmental regulations even in the absence of a government enforcement threat. Foreign firms, if anything, help fill the China's implementation gap.

The remainder of this chapter addresses this debate by exploring the environmental behavior of foreign firms as well as their relationship with their Chinese supply partners. It argues that, while multinational enterprises can certainly exhibit a greater level of environmental responsibility in their China operations, overall they can contribute constructively to China's environmental governance. However, the extent to which they do so depends on an array of other factors related to the firm, industrial sector, and host jurisdiction.

2. OUTBREAK: A RASH OF POLLUTION CASES INVOLVING FOREIGN FIRMS

Scanning the headlines over the last few years would certainly lead one to conclude that foreign firms are a core cause of China's pollution problem. Dozens of foreign companies have been caught violating China's environmental regulations and in several cases generated a firestorm of criticism. In 2009, Greenpeace released a report showing that eight multinationals not only exceeded discharge standards, but subsequently failed to adhere to disclosure regulations (Greenpeace 2009). In 2011, American energy company ConocoPhillips was widely condemned in the Chinese media for its role in oil spills that affected approximately 2,400 square miles in the Bohai Sea. Following the spill, and perceived lack of transparency of ConocoPhillips, one official from the State Oceanic Administration lamented, "China is weak in supervision over marine development, especially the offshore oil exploration operated by multinational corporations" (Wang 2011). Noting that ConocoPhillips faced a rather modest maximum fine of $31,625, a *People's Daily* editor accused it of finding "profitable loopholes in China's environmental law and system" (Liu 2011). The *Beijing Evening News* (2011) reproached ConocoPhillips for abandoning "the basic principle of honesty." The *Global Times* (2011) saw the incident as indicative of a larger issue arguing, "ConocoPhillips is like certain Westerners who live in China. They can be quite law-abiding back home, but break the law in China. Some global corporations can even outsmart their Chinese counterparts in under-the-table deals, bribing officials and cutting employee benefits." ConocoPhillips eventually paid RMB 1.2 billion in compensation for the environmental damage, while its Chinese partner CNOOC (China National Offshore Oil Corporation) paid another RMB 480 million.

Just as the ConocoPhillips's case was heating up in late 2011, authorities in Chongqing temporarily shut down all 13 of the city's Walmart stores and fined the company $420,000 for selling pork falsely labeled as organic. Around the same time, Shanghai authorities accused a battery-making unit of American multinational Johnson Controls of sickening 49 local children through excessive discharge of lead into local waterways. The Johnson factory, along with others in the area, was shuttered, although in the end it simply moved its lead-acid manufacturing to other Chinese facilities (Metal Bulletin 2012). In 2012, demonstrations in Qidong led the government to cancel the construction of a wastewater drainpipe project that was part of a Japanese paper factory (BBC 2012).

Several dozen multinationals have been charged with contributing to China's pollution problem through inadequate supervision of their supply chains. Starting in April, 2010, the Institute for Public and Environmental Affairs (IPE) ran a series of reports on heavy metal pollution discharged by manufacturing firms in the information technology (IT) sector. The firms highlighted in the report were suppliers for 29 well-known brands, the overwhelming majority of which were foreign companies such as Apple,

Panasonic, Samsung, and IBM. The reports detailed a host of violations by the suppliers such as exceeding wastewater discharge limits, failure to properly handle hazardous waste, and neglecting to keep or disclose environmental records.[3]

Apple, in particular, came under intense scrutiny. Apple's global prominence and extensive supplier network in China, combined with its culture of secrecy and initial unwillingness to engage its critics, opened it up for particularly scathing criticism both in China and the West.[4] In its initial report, IPE described in detail how eight of Apple's suppliers were endangering their workers and/or violating pollution regulations (Friends of Nature; Institute for Public and Environmental Affairs; Green Beagle 2011). Following Apple's failure to respond to IPE's criticisms, IPE issued a second report that detailed violations by additional suppliers and concluded: "Apple has already made a choice; to stand on the wrong side, to take advantage of the loopholes in developing countries' environmental management systems, and to be closely associated with polluting factories so that it can continue to grab their own super profits, at the expense of the environment and communities; becoming a barrier in China's path towards pollution reduction" (Friends of Nature et al. 2011).

The textile and apparel industry has also been a target of environmental activists. Greenpeace produced a series of reports that, much like the IPE investigations of the IT sector, focused on the pollution discharges of suppliers for famous international brands such as Calvin Klein, Converse, and Lacoste. In its two series, entitled "Toxic Threads" and "Dirty Laundry," Greenpeace gathered water samples outside waste water treatment plants serving industrials zones that are largely populated by textile firms. While there was no indication that these chemicals were the result of transgressions of environmental regulations,[5] the samples contained a number of chemicals (e.g., chlorinated benzenes) deemed hazardous both to the ecology and human health. IPE also produced a report on the textile industry, which uncovered a number of air and water pollution discharge violations among suppliers for the approximately four dozen international brands on which it focused. The list included companies like H&M, Disney, J.C. Penney, and Ralph Lauren (Ma et al. 2012).

3. THE MISBEHAVING MULTINATIONAL: TREND OR OUTLIER?

It is evident that there has been no shortage of cases in which the polluting activities of foreign companies have drawn the attention of government regulators, media, and general public. The question is: how do we interpret these cases? What does the wave of pollution incidents involving foreign-invested enterprises tell us about their overall impact on China's environmental regulations? Are these cases indicative of a wider trend of foreign firms systematically exploiting China's environment and turning it into a pollution haven? Or, are these examples simply anomalous incidents that are inevitable in a country with almost half a million FIEs?

The answer lies somewhere in between the two competing portraits of foreign investors. Foreign investors are neither wanton polluters nor are they environmental stewards. Openness to foreign investment can help narrow China's implementation gap, but FDI's impact is largely conditional and depends on factors such as the firm's country of origin, industrial sector, as well as features of the host jurisdiction such as the strength of local civil society. The remainder of this chapter forwards this argument. I start by questioning the claims of those who tend toward the optimistic end of the spectrum and see a green dividend from China's reliance on FDI. Subsequently, I refute the pollution haven argument that sees foreign investors as primarily exploiting the weakness of China's environmental regulations.

The steady drumbeat of pollution cases involving foreign investors casts doubt on the assertions of those that make the strongest case for the environmental benefits of economic globalization. It is apparent that a significant number of foreign-invested companies are not engaging in rigorous self-regulation and that foreign investment is not necessarily "greening China" (Zeng and Eastin 2011). Foreign companies regularly appear on the pollution database map maintained by IPE. When an environmental NGO examines a particular sector in-depth, such as the electronics business, many of the industry leaders appear linked to pollution violations through their suppliers. Foreign-invested companies from Hong Kong and Taiwan are repeatedly found breaching environmental regulations (Yu and Wong 2011:81).

And even if most foreign companies do not run afoul of regulators, many appear largely willing to comply with the law of the land rather than expend resources to promote environmental protection. MNCs are not "exporting environmentalism" as extensively as some depicted (Garcia-Johnson 2000). There is truth to the claim, frequently heard in China, that the environmental behavior of many foreign companies exhibits a kind of double standard in that their environmental performance does not match the lofty rhetoric of their corporate environmental statements (Yoshioka 2012). This is evident in the reports published by IPE and Greenpeace, which time and again show MNCs adopting a *de facto* "don't ask, don't tell" attitude when it comes to the environmental practices of suppliers. This MNC "head in the sand" mentality is particularly the case for Chinese partners that are not direct, or first-tier, suppliers. One of the more interesting aspects of the IPE and Greenpeace reports, which often summarize the NGO's correspondence with the MNCs, is the reaction of many multinationals when confronted with evidence that their suppliers have a history of environmental violations. Many questioned whether a particular violator was in their supply chain; others indicated that they did not know the company had committed an infraction or simply noted that the company was a second-tier supplier (and therefore presumably not the responsibility of the multinational). The overall impression is that MNCs' requests to their suppliers to adhere to environmental regulations are often *pro forma* and not aimed at exerting pressure on suppliers' environmental management (Lam 2011).

By contrast, MNCs' demands for high-speed delivery and low price exert a significant influence on their supply chain. Unfortunately, it is one that is often at odds with environmental protection. As Ma Jun, winner of the Goldman Environmental Prize and founder of IPE, states, "For foreign companies, China is all about price... They lure local suppliers to cut corners to win contracts and that is very negative for China's environment." (Ford 2012). MNCs may request that their suppliers spend more on pollution abatement, but the investment is typically expected to come at the suppliers' expense. This means lower profits for suppliers that already feel as if they are operating on thin margins, and presumably are not predisposed toward strong environmental management. Thus, it is hardly surprising that the requests are either ignored or Chinese suppliers hide their real production facilities from MNC auditors (Harney 2008, esp. Ch. 2). For example, it is unlikely that Apple ever directly authorized its Chinese supplier to use N-hexane, a chemical used to clean iPhone screens that poisoned dozens of workers. But it is easy to imagine that the chemical's use was at least in part a function of production pressure. N-hexane dissolves three times faster than traditional alcohol-based cleaners, which obviously matters for a supplier operating on a tight margin and asked to produce thousands of iPhones per day (Duhigg and Barbosa 2012).

One sees a similar story with Walmart, which has enjoyed a public relations boost for its efforts to green its supply chain. Starting in 2008, Walmart announced goals such as eliminating 20 million tons of greenhouse gas emissions from its global supply chain, working with its top 200 suppliers in China to improve energy efficiency by 20 percent (per unit of production), and requiring suppliers to complete sustainability assessments. Toward those latter goals, Walmart teamed up with NGOs in China such as Environmental Defense Fund, Natural Resources Defense Council, as well as non-profit organizations such as the China Technical Institute (CTI).[6] Between 2008 and 2011, Walmart representatives and NGO partners conducted more than 300 factory visits in China. While this is clearly a laudatory initiative with the potential to exert a profound influence across China, 300 factory visits is a drop in the bucket considering Walmart procures from more than 10,000 Chinese suppliers (Ma et al. 2010). And by some accounts Walmart's efforts to green the supply chain have been undercut by its notorious frugality and price pressure on suppliers. Walmart representatives have been periodically pulling out of the project, leaving the work to NGO partners (Plambeck and Denend 2011:21). Some reports conclude that Walmart's initiatives in China have been largely superficial, lacking follow-up, and marginally successful at best. One investigative report quotes a Chinese partner of Walmart as stating, "They will work suppliers to the death" and points out that "operating on the thinnest of margins and scrambling to keep up with Walmart's demands... factories just don't have the time or capital to invest in green projects" (Kroll 2012).

But before one starts accusing foreign investment of turning China into a pollution haven, it should be pointed out that, despite the headlines, the number of foreign firms that violate China's pollution laws is relatively small, both in comparison to domestic companies and as a percentage of all foreign firms in China. Even if one accounts for the fact that there are far less foreign than domestic firms in China, FIEs are underrepresented on lists of industrial polluters. IPE's initial list of firms violating discharge standards included 33 multinationals, but more than 2,600 Chinese firms (Zhang 2006).[7] And it is widely accepted that foreign companies, perceived by regulators as deep-pocketed and therefore readily capable of adhering to environmental standards, are often singled out for enforcement activity (Gerdes 2011). Scholars have produced many large-sample, firm-level surveys focusing on the drivers of corporate environmental performance, yet to my knowledge no study has found that foreign-ownership is associated with non-compliance. On the contrary, studies have found that under certain conditions, foreign ownership leads to better environmental practices.

For instance, Stalley (2010) conducted a survey of corporate environmental behavior using the China Green Watch program. This program is a joint initiative between Chinese environmental authorities and the World Bank that ranks firms on a five-scale metric (green, blue, yellow, red, and black). Rankings are based on a variety of criteria including the company's compliance history and internal environmental management policies. Drawing on a survey of 228 firms in three cities, and controlling for other factors associated with corporate environmental behavior (e.g., profitability), Stalley finds that foreign ownership is not correlated with a firm's Green Watch ranking (in a statistically significant manner). In other words, Stalley finds no evidence that foreign firms outperform Chinese firms in terms of environmental management. However, Stalley does find that companies that are foreign-controlled, that is firms in which the foreign investor owns a majority stake, present superior environmental performance. Furthermore, foreign firms in China with investment from OECD countries or Japan are particularly prone to strong environmental management. Of the European firms in his sample, almost 70 percent achieve the highest possible Green Watch ranking. Zeng and Eastin (2011), who perform a survey of business executives in China, reach a similar conclusion about the high performance of FIEs from OECD countries. The reason for the superior performance of developed country firms is straightforward—they typically possess sufficiently advanced technology and refined environmental management practices so that compliance with China's discharge standards is easy. The savings gained from shirking environmental regulations are too modest to warrant engaging in illegal activity (He 2006:229).

These findings tell us two things. First, there is little support for the idea that foreign companies are systematically exploiting China in the manner anticipated by the harshest critics of economic globalization. There is simply no evidence indicating that foreign companies violate China's environmental standards with greater frequency than do domestic firms. Indeed, 65 percent of the foreign enterprises in Stalley's sample score "blue" or "green," both of which indicate the company has above-standard performance. MNCs may not live up to their environmental responsibility with regard to supervising their suppliers, but they tend to assure their own facilities meet Chinese standards. At the same time, however, Stalley's findings cast doubt on the rosier assessments of the pro-globalizationists that see a broad environmental benefit from China's opening to foreign investment. As indicated above, the environmental influence of foreign investment is conditional. It depends on whether the foreign investor takes a controlling share and/ or comes from Europe and Japan. Since almost half of China's FDI comes from Hong Kong, much of China's FDI may exert no discernible positive influence on China's environmental regulations. Under the right conditions, foreign investment may narrow China's implementation gap as some FIEs go beyond compliance, but it is not simply the case that more foreign investment means better environmental protection.

Moving away from firm-level scholarship and focusing instead on cross-regional studies, the story is much the same. Again, one sees little support for the idea that foreign investors undermine China's environmental regulations. Very few studies have found evidence that foreign investment is drawn to jurisdictions with poor implementation of environmental standards. Looking at Chinese provinces from 1996 to 2004, Zeng and Eastin (2011) conduct a quantitative analysis of the impact of a province's pollution levy on foreign direct investment. The pollution levy serves as a measure of the stringency of environmental regulations; the higher the average levy paid per firm the more stringent the enforcement. Controlling for factors influencing a province's ability to attract FDI, such as quality of the labor force and rate of economic growth, they find that FDI does not flow to areas with weak environmental regulations. If anything, FDI is attracted to more environmentally stringent locations. Dean, Lovely, and Wang (2009) look at data for Chinese provinces between 1993 and 1996 and reach a similar finding that, as a whole, foreign investment does not seek out weak environmental regulations. However, they do find that foreign investment from Hong Kong, Macao, and Taiwan in pollution-intensive sectors is drawn to weak environmental standards. Again, the conclusion is that the impact of foreign investment, whether harmful or constructive, is contingent upon other factors, particularly the country of origin of the investor.

The role of foreign companies in the process of transfer of clean technologies also cannot be overlooked. As MNCs transfer advanced technology to subsidiaries and joint venture partners, technological spillover should ultimately reduce the cost of compliance with discharge standards for both domestic and foreign firms. In that sense, foreign investment narrows the implementation gap. Some of this technological transfer takes place via foreign investment directed through the Clean Development Mechanism (CDM).[8] The global investment of all CDM projects is over $215 billion with the lion's share going to China and (to a lesser extent) India. The two countries account for 65 percent of all investment (Kirkman et al. 2012:8). There are more than 1,800 registered CDM projects in China, and although CDM-related foreign investment represents a small fraction of China's overall FDI, it contributes to the transfer of advanced technology (both equipment and knowledge).[9] One study found that 59 percent of CDM projects included some form of technology transfer and that China's average was far higher than other large developing countries such as India (12 percent) and Brazil (40 percent) (Dechezlepretre, Glachant and Meniere 2009).[10]

Aside from CDM, in recent years foreign investment has made positive contributions to China's acquisition of clean technologies. This was not always the case. In the early reform period, China struggled to prevent foreign investors from transferring and utilizing "backwards" technologies that were often highly polluting and in some instance banned in developed countries. In the absence of incentives provided by Chinese government regulations, foreign investors were reluctant to transfer more advanced technologies. Gallagher's (2006) study of the auto industry found that US auto manufacturers transferred relatively ineffective emission control technologies to joint venture partners. The technology was sufficient to meet Chinese standards, but would not comply with US air pollution requirements. Starting in 1995, however, China began issuing foreign investment guidance catalogues, the most recent of which was issued in 2011. These catalogues offer a list of sectors where foreign investment is encouraged, restricted, and prohibited; some of the prohibited sectors included those deemed excessively polluting. Also starting in the mid-1990s, Beijing issued a separate set of catalogues specifying pollution-intensive technologies and processes that were to be eliminated (and not approved for foreign investment). Collectively, these catalogues have contributed to the reduction of highly polluting foreign investment projects. Over the last decade China has also been actively pursuing additional policies that contribute to the transfer of clean technologies. For instance, local content requirements for foreign-invested wind power projects have helped transfer advanced technologies to China (Lewis 2013). A study of China's power sector found that FDI contributed positively to China's environment energy efficiency. FDI projects were below the government's heat rate targets, a measure of energy efficiency. The foreign projects were also more efficient than new Chinese projects and on par with similar-sized plants in the United States (Blackman and Wu 1999:707). In short, by serving as a conduit for the introduction of advanced, environmentally friendly technology into China, foreign investment reduces the pollution intensity of industry and facilitates compliance with environmental regulations.

Finally, it bears mentioning that there is evidence that the beneficial impacts of foreign investment may be gradually broadening beyond a small number of companies from developed nations. In particular, environmental activists' pressure is pushing MNCs to invest greater effort in working with their Chinese partners in the area of environmental management. Over the last decade, it has become common practice for the largest MNCs to conduct green supply chain (GSC) policies. Usually, this has involved an MNC reviewing potential suppliers' environmental management practices and periodically conducting audits of the facilities operations. In the past, the impact of multinationals' green supply chain policies on their Chinese partners has been relatively limited.[11] Some surveys have found evidence that GSCs have influenced the environmental management practices of Chinese suppliers. For instance, Zhu and Sarkis (2006) argued that Chinese suppliers in the auto industry improved their environmental performance in order to access foreign customers in China. But this is largely a function of the nature of the automobile sector, where long-term commercial partnerships are common. More often scholars have not discovered evidence of a widespread GSC effect. Stalley (2010), for instance, found that whether a Chinese firm possessed a commercial relationship with an MNC in China had no statistically significant impact on its environmental performance. The reasons for this are straightforward. First, only the largest, highest-profile MNCs conduct green supply chain policies. The vast majority of foreign companies in China have little to no policy of supervising commercial partners. Second, as discussed above, many MNCs have traditionally conducted their GSC practices in a narrow fashion, focusing only on a few top suppliers or on Chinese partners whose activities represent the greatest reputational risk to the MNC (e.g., a Chinese firm that is contracted to transport hazardous chemicals produced by the MNC).

But this situation is gradually changing. Most importantly, a combination of an increasing NIMBY (not-in-my-backyard) mentality among Chinese citizens along with the persistent efforts of environmental activists to shame polluting companies is putting much more pressure on MNCs and, by extension, their Chinese suppliers. Five hundred seventy companies have approached IPE seeking to get off its list of polluters largely based on their fear of losing contracts with foreign companies (Ford 2012). After more than a year of intense pressure, in 2012 Apple released the names of 156 of its suppliers, admitted that more than a dozen factories had violated pollution laws, and indicated it was suspending business with those known to have violated regulations while requiring others to undergo third-party audits (Ford 2012; Yoshioka 2012). At the same time, it tripled its corporate social responsibility staff (Bradsher and Duhigg 2012). Many other big brand names, such as Panasonic and Philips, have indicated that they will supervise their suppliers' environmental record using the IPE pollution database. Several dozen consumer product MNCs, such as Walmart, Gap, and Adidas, work with CTI, which has trained more than 8,000 managers in areas related to business sustainability. And according to a representative of CTI, there is the beginning of a movement away from the top-down "audit-based" approach in favor of a more "partnership-based" relationship between foreign buyers and Chinese suppliers.[12] What this means is that MNCs are going beyond the old model in which they simply issued requests for environmental compliance and required suppliers to undergo the occasional inspection, but largely remained hands-off and offered very limited support. Now, leading MNCs are partnering with suppliers and third parties to perform training, identify methods and technologies for waste reduction, build capacity, and more broadly enhance environmental awareness. Organizations such as CTI are critical in this transition in both providing consulting services to MNCs and, perhaps more importantly, making Chinese suppliers aware of the business benefits of stronger environmental management.

4. Conclusion

This chapter has explored the role of foreign companies in China's environmental regulations. It began by pointing out that China faces a considerable pollution challenge, which is largely driven by industry's lack of compliance with pollution control regulations. At the same time, thousands of overseas investors are pouring into China each year and foreign companies occupy a prominent position in China's economy. The implication is obvious: if China is to address its environmental dilemma successfully, it needs the active cooperation of foreign investors.

A recent spate of high-profile pollution incidents involving multinationals, such as ConocoPhillips's oil spill in the Bohai Sea, leaves an impression that foreign investors are despoiling China's environment and taking advantage of its weak environmental regulations. But, as I argued, these cases belie the general trend. As a rule, foreign investors comply with China's environmental regulations. There is almost no scholarship that indicates foreign companies consistently perform below China's standards or that their pollution control is inferior to domestic firms'. Moreover, foreign investment can often serve to boost compliance through the transfer of clean technologies and the imposition of green supply chain policies in which foreign companies attempt to improve the environmental management practices of their Chinese partners.

However, one cannot take the positive portrayal of foreign investment too far. Foreign firms in China, like companies all over the world, care more about profits than environmental protection. In the absence of external oversight from government regulators and NGOs, it is unlikely foreign companies will do much beyond meeting the basic requirements of Chinese law. Multinationals have shown modest enthusiasm for expending resources to turn their own lofty pledges about environmental and social responsibility into reality. Foreign investment, or more broadly economic openness, is not itself a solution to China's environmental dilemma. While foreign companies *can* green China, whether they actually *do* so depends on a host of factors. As I have argued throughout this chapter, the impact of foreign investment is contingent.

To start with the obvious, government regulation is critical. Foreign investors proved reluctant to transfer advanced, clean technology until Beijing began outlawing dirty manufacturing processes and issuing regulations such as local content requirements. Also, various features of the firm, sector, and host destination each help influence the extent to which foreign investors play a constructive role in environmental regulations. All things being equal, such FIEs in which the foreign investor owns a controlling share and is from an OECD country are most likely to go beyond compliance in their own facilities. Most multinationals perform some oversight of their suppliers' environmental management, but as the Walmart example demonstrates, certain features of the market influence the extent of MNC influence. For price-intensive sectors, where the cost of the manufactured item matters more than its brand name, companies appear less likely to implement robust green supply chain policies. By contrast, sectors in which long-term relationships between foreign buyers and Chinese suppliers are common, such as the auto industry, can increase the effectiveness of green supply chain policies. And finally, one cannot underestimate the importance of civil society organizations such as IPE and Greenpeace. As the case of Apple shows, when a company or sector is on the radar of corporate watchdogs, it is far more likely to fulfill its environmental responsibility. Ultimately, the development of civil society represents a more pressing need for China's environmental protection than does the expansion of foreign investment.

Notes

1. For a more extended discussion of the pollution haven debate, see Stalley (2010).
2. In this chapter, I use both the term foreign-invested enterprise (FIE) and multinational corporation (MNC). The former refers to any company in China that has received foreign investment, while the latter term refers to companies that have assets in China (e.g., factories, offices), but are headquartered elsewhere. Also, when referring to MNCs I am typically referring to large multinationals such as those that occupy the Fortune 500 list. With regard to China's enforcement gap, in this chapter my main focus is discharge standards for air and water pollution. I do not discuss other environmental regulations relevant to business such as the requirement that any industrial project must complete and obtain approval for an environmental impact assessment, adhere to the "three simultaneous," etc.
3. These reports are available on IPE's website: http://www.ipe.org.cn/En/about/report.aspx.
4. Simply the title of Christina Larson's (2011) *Foreign* Policy piece about Apply in China, "Red, Delicious, and Rotten," provides a clear sense of the criticism directed at Apple in the West.
5. Indeed, as the Greenpeace report makes clear, at the time there was no regulation covering toxic chemicals (Greenpeace 2012b:18).
6. CTI is an independent organization in Guangzhou established in 2004 with State Department financial assistance by BSR, a global non-profit member-based organization that provides consulting in a variety of areas related to corporate sustainability.
7. Unfortunately, IPE does not keep data on the ownership structure of the thousands of companies that are listed in its database of pollution violators, so at this point in time it is difficult to determine the precise ratio of foreign to domestic firms.
8. CDM is a carbon market created as part of the 1998 Kyoto Protocol that allows developed countries to earn credits through investment in emission reduction projects in developing countries.

9. About 6 percent of China's CDM projects involve foreign investment (Kirkman et al. 2012:50).
10. Kirkman et al. (2012:30) put China's average technology transfer rate much lower at 20 percent. In more recent years, it is closer to 10 percent. China's declining rate is typical. As more projects are established, there is less transfer because earlier projects provide local technology access. In addition, technology access is facilitated by other channels such as non-CDM foreign direct investment.
11. There is evidence that concerns about "green trade barriers" has pushed Chinese exporters, especially those who rely heavily on European and North American markets, to pursue "beyond compliance" environmental management policies such as ISO14000 certification. However, the focus of this chapter is not on the influence of exporting, but rather on the influence of foreign investment in China. For more on the impact of exporting, see Stalley (2009).
12. Author's interview with Jason Ho, Manager, Advisory Services and CTI, BSR, March 31, 2013.

Bibliography

BBC. 2012. "Japanese Pollution Discharge Project in China To Be Cancelled." *BBC Worldwide Monitoring*, July 28.
——. 2013. "China Acknowledges 'Cancer Villages'." *BBC*, February 22.
Beijing Evening News. 2011. "Oil Giant Must Pay Price For False Words." *Global Times*, September 4.
Blackman, Allen and Xun Wu. 1999. "Foreign Direct Investment in China's Power Sector: Trends, Benefits, and Barriers." *Energy Policy* 27: 695–711.
Bloomberg. 2011. "Half Groundwater of 182 China Cities Not Drinkable, Daily Says." *Business Week*, October 20.
Bradsher, Keith and Charles Duhigg. 2012. "Signs of Change Taking Hold in Electronics Factories in China." *New York Times*, December 27.
Buckley, Chris. 2011. "Heavy Metals Pollute a Tenth of China's Farmland." *Reuters*, November 6.
Dean, Judith M., Mary E. Lovely and Hua Wang. 2009. "Are Foreign Investors Attracted to Weak Environmental Regulations? Evaluating the Evidence from China." *Journal of Development Economics* 90: 1–13.
Dechezlepretre, Antoine, Matthieu Glachant and Yann Meniere. 2009. "Technology Transfer By CDM Projects: A Comparison of Brazil, China, India and Mexico." *Energy Policy* 37: 703–711.
Duhigg, Charles and David Barbosa. 2012. "In China, Human Costs Are Built Into an iPad." *New York Times*, January 25.
Economy, Elizabeth. 2007. "The Great Leap Backward?" *Foreign Affairs* 86(5): 38–59.
Feng, Jie. 2013. "China's Groundwater Scandal: The Missing Waste Water." *China Dialogue*. March 4. Accessed March 10, 2013. http://www.chinadialogue.net/article/show/single/en/5764-China-s-groundwater-scandal-the-missing-wastewater.
Ford, Peter. 2012. "Ma Jun Helps Chinese Find Out Who's Pollution and Shame Corporations Into Cleaning Up." *Christian Science Monitor*, April 16.

Friends of Nature, Institute of Public and Environmental Affairs, and Green Beagle. 2010. *The IT Industry Has a Critical Duty to Prevent Heavy Metal Pollution.* Beijing: Institute of Public and Environmental Affairs.

———. 2011. *The Other Side of Apple.* Beijing: Institute for Public and Environmental Affairs.

Friends of Nature; Institute of Public and Environmental Affairs; Green Beagle; Envirofriends; Green Stone Environmental Action Network. 2011. *The Other Side of Apple II.* Beijing: Institute of Public and Environmental Affairs.

Gallagher, Kelly Sims. 2006."Limits to Leapfrogging In Energy Technologies? Evidence From The Chinese Automobile Industry." *Energy Policy* 34: 383–394.

Garcia-Johnson, Ronie. 2000. *Exporting Environmentalism: U.S. Multinational Chemical Corporations in Brazil and Mexico.* Cambridge, MA: MIT Press.

Gerdes, Justin. 2011. "Violators Beware." *China Dialogue*, December 13. Accessed April 2, 2013. http://www.chinadialogue.net.

Global Times. 2011. "ConocoPhillips Faces an Ethical Test." *Global Times*, August 25.

Greenpeace. 2009. *Silent Giants: An Investigation Into Corporate Environmental Information Disclosure In China.* Beijing: Greanpeace.

———. 2011. *Dirty Laundry: Unravelling the Corporate Connections to Toxic Water Pollution in China.* Amsterdam: Greenpeace International.

———. 2012a. *Dangerous Breathing.* Beijing: Greenpeace China.

———. 2012b. *Toxic Threads: Putting Pollution on Parade.* Amsterdam: Greenpeace International.

Harney, Alexandra. 2008. *The China Price: The True Cost of Chinese Competitive Advantage.* New York: Penguin Press.

He, Jie. 2006. "Pollution Haven Hypothesis and Environmental Impacts of Foreign Direct Investment: The Case of Industrial Emission of Sulfur Dioxide (SO2) in Chinese Provinces." *Ecological Economics* 60: 228–245.

Hook, Leslie. 2013. "Scientist Hits Out Over China Air Quality." *Financial Times*, January 17.

Huang, Yanzhong. 2013. "Choking to Death: Health Consequences of Air Pollution in China." *Asia Unbound, Council on Foreign Relations.* March 4. Accessed March 5, 2013. http://blogs.cfr.org/asia/2013/03/04/choking-to-death-health-consequences-of-air-pollution-in-china/#more-10777.

Kirkman, Grant A., Stephen Seres, Erik Haites and Randall Spalding-Fecher. 2012. *Benefits of the Clean Development Mechanism 2012.* Bonn, Germany: United Nations Framework Convention on Climate Change.

Kroll, Andy. 2012. "Are Walmarts Chinese Factories as Bad as Apple's." *Mother Jones*, March 21.

Lam, Maria Lai-Ling. 2011. "Challenges of Sustainable Environmental Programs of Foreign Multinational Enterprises in China." *Management Research Review* 34(11): 1153–1168.

Larson, Christina. 2011. "Red, Delicious, and Rotten: How Apple Conquered China and Learned to Think Like the Communist Party." *Foreign Policy*, August 1.

Lewis, Joanna. 2013. *Green Innovation in China: China's Wind Power Industry and the Global Transition to a Low-Carbon Economy.* New York: Columbia University Press.

Lim, Louisa. 2013. "Beijing's 'Airpocalypse' Spurs Pollution Controls, Public Pressure." *National Public Radio.* January 14. Accessed March 2, 2013. http://www.npr.org/2013/01/14/169305324/beijings-air-quality-reaches-hazardous-levels.

Liu, Rui. 2011. "Pollution Oil Companies Not Above the Law." *Global Times*, August 22.
Lo, Carlos Wing-Hung, Gerald E. Fryxell, Benjamin van Rooij, Wei Wang and Pansy Honying Li. 2012. "Explaining the Enforcement Gap in China: Local Government Support and Internal Agency Obstacles as Predictors of Enforcement Actions in Guangzhou." *Journal of Environmental Management* 111: 227–235.
Ma, Jun, Jingjing Wang, Matthew Collins, Malei Wu, Sabrina Orlins and Jie Li. 2012. *Sustainable Apparel's Critical Blind Spot*. Beijing: Friends of Nature, Institute of Public and Environmental Affairs, Green Beagle, Envirofriends, and Nanjing Green Stone.
Ma, Jun, Ray Cheung, Jingjing Wang and Qingyuan Ruan. 2010. *Greening Supply Chains in China: Practical Lessons From China-based Suppliers in Achieving Environmental Performance*. Washington DC: World Resources Institue.
Metal Bulletin. 2012. "Johnson Controls Stops Lead Processing in Shanghai." *Metal Bulletin Weekly*, September 28.
Morrison, Wayne. 2013. *China's Economic Rise: History, Trends, Challenges, and Implications for the United States*. Washington DC: Congressional Research Service.
OECD. 2007. *OECD Environmental Performance Review*. Paris: Organization for Economic Cooperation and Development.
Plambeck, Erica and Lyn Denend. 2011. "The Greening of Walmart's Supply Chain... Revisited." *Supply Chain Management Review* September/ October: 16–23.
Ross, John. 2012. "Why FDI Into China Outperformed the World." *China.org.cn*. December 24. Accessed April 1, 2013. http://www.china.org.cn/opinion/2012-12/24/content_27500147.htm.
Stalley, Phillip. 2009. "Can Trade Green China? Participation in the Global Economy and the Environmental Performance of Chinese Firms." *Journal of Contemporary China* 18(61): 567–590.
——. 2010. *Foreign Firms, Investment, and Environmental Regulation in the People's Republic of China*. Stanford: Stanford University Press.
Tilt, Bryan. 2007. "The Political Ecology of Pollution Enforcement in China: A Case from Sichuan's Rural Industrial Sector." *China Quarterly* 192: 915–932.
Waldmeir, Patti. 2013. "China Pollution: Fears Over Poor Air Exacerbate Healthcare Concerns." *Financial Times*, March 1.
Wang, Aileen and Nick Edwards. 2012. "China 2012 FDI Inflows Slow, Stay On Track For $100 Billion." *Reuters*, November 19.
Wang, Qing. 2011. "ConocoPhillips Urged to Apologize For Oil Leak." *China Daily*, August 12.
Warwick, Mara. 2003. "Environmental Information Collection and Enforcement at Small-Scale Enterprises in Shanghai: The Role of Bureaucracy, Legislatures, and Citizens." PhD diss., Stanford University.
Watts, Jonathan. 2012. "The Clean-up Begins on China's Dirty Secret—Soil Pollution." *The Guardian*, June 12.
Wong, Edward. 2013a. "Air Pollution Linked to 1.2 Million Premature Deaths in China." *New York Times*, April 2.
——. 2013b. "Spill in China Underlines Environmental Concerns." *New York Times*, March 3.
World Bank. 2007. *Cost of Pollution in China: Economic Estimates of Physical Damage*. Washington DC: World Bank.
Xinhua. 2008. "Most of World's Top Companies Invest in China." *Xinhua*, May 5.

———. 2012. "China's 2011 FDI Hits 116.01 Bln USD." *Xinhua*, January 18.
Xu, Qi. "Facing Up to 2007. 'Invisible Pollution'." *China Diaologue*, January 29. Accessed March 2, 2013. http://www.chinadialogue.net/author/121-Xu-Qi.
Yoshioka, Keiko. 2012. "INTERVIEW/ Ma Jun: Is China Ready for Greening?" *Asahi Shimbun*, July 12.
Yu, Fu. 2013. "Air Quality Poor in 90 percent of Chinese Cities." *China Radio International*. March 29. Accessed March 29, 2013. http://english.cri.cn/6909/2013/03/29/2941s756703.htm.
Yu, Xiaojiang and Koon Kwai Wong. 2011. "Environmental Performance of Foreign Direct Investment (FDI) Companies in the Pearl River Delta Region (PRDR): A Case Study of Dongguan City." *Australian Geographer* 42(1): 79–93.
Zeng, Ka and Joshua Eastin. 2011. *Greening China: The Benefits of Trade and Foreign Direct Investment*. Ann Arbor: University of Michigan Press.
Zhang, Chun. 2013. "Why Should Soil Pollution Data Be Kept a 'State Secret'?" *China Dialogue*. March 4. Accessed March 5, 2013. http://www.chinadialogue.net/article/show/single/en/5758-Why-should-soil-pollution-data-be-kept-a-state-secret-
Zhang, Liuhao. 2006. "Blacklist Marks Foreign Offenders." *Shanghai Daily*, October 27.
Zhu, Qinghua and Joseph Sarkis. 2006. "An Inter-sectoral Comparison of Green Supply Chain Management in China: Drivers and Practices." *Journal of Cleaner Production* 14: 472–486.

PART III

ENVIRONMENTAL MOVEMENTS AND PUBLIC PARTICIPATION

CHAPTER 7

ENVIRONMENTAL PROTESTS AND LOCAL GOVERNANCE IN RURAL CHINA

Bingqiang Ren

1. INTRODUCTION

China's social transition in recent years has been characterized by the proliferation of social conflicts and unrests as social interests have diversified and yet the effective mechanisms to mitigate the conflict are lacking. In rural China, the number of petition and protests over land disputes, elections, pollutions, among many other issues, has increased drastically in recent years. More worrisome is the tendency of large-scale violent protests seen in many conflicts in various locations.

Scholarly research on rural protests is not new. The existing literature, however, suffers from two problems. First, the existing studies tend to differentiate peasant movements into two types that are not inherently connected. The first type is what O'Brien and Li (2006, also O'Brien 2009) labeled the "rightful resistance," the action within the legal boundary, such as petition. This becomes an interesting research subject because in many cases villagers deliberately avoided confrontation with the authority through measures such as protests and riots but instead strategically utilized the existing policies and regulations to express their grievances and voice their demands. It is disputed among scholars whether this "rightful resistance" represents a genuine consciousness of rights among peasants. O'Brien and Li argue that villagers in this "policy-based resistance" sought policy adjustment in order to meet their immediate interests rather than aiming to defend their rights *per se*. Yu (2004), on the other hand, argues that many cases of "rightful resistance" suggested a budding stage of civil movements with right-conscious leaders and a well-organized platform. Regardless, this line of research considers "rightful resistance" different from the so-called mass protests that fall outside the legal boundary defined by the Chinese government. Ying (2007), for example, differentiates the concepts of "mass interest express mechanism" (e.g., petition) and "mass protests" and argues that the two types of movements differ fundamentally due to the different attitudes toward legal constraints.

Such separation, however, is problematic because it ignores the inherent connection between the two types of actions. This chapter argues that the two types of actions, though one is legal and the other illegal, often occur sequentially and both can be considered as political participation through which individuals seek policy revision in order to defend their interests. What matters is what motivates individuals to choose one action over another under specific circumstances.

The "specific circumstances" speak to the second problem in the current research, which overwhelmingly focuses on the internal logic of collective action among peasants and its impact on the outcomes of peasant movements. Much research examines factors such as relative deprivation, leadership, framing, etc. that consume much energy of the social movements literature. What is overlooked in the literature, however, is that the actions peasants take, whether legal or illegal, are embedded in local politics and the systems where interaction between villagers and governmental officials takes place. It is, therefore, imperative to recognize that local governments' behavior and the political system in which it takes place have deep influence on the nature and form of mass political participation among rural residents.

Although peasant protests occur with various causes, environmental protests have become a prominent form. Along with China's rapid industrialization, pollution has become a serious threat to many rural residents. The conflict associated with environmental deterioration rises to be a new force affecting China's social stability. This chapter attempts to understand the mass movements caused by environmental conflict. It will pay particular attention to the process in which these movements transformed from peaceful negotiation to violent protests and the particular role local governments played in inciting this transformation. It will argue that the key factor in this transformation is the deficiency of governance in local governments.

2. Description of Four Cases

Four cases of environmental protests are discussed and analyzed in this study. The cases are selected based on two considerations. First, availability of information on protests is always a problem in China for a meaningful analysis. New media often offer very brief sketch for most cases due to either the scale of the event or most importantly the censorship on reports of protests. It is therefore not easy to find detailed description about a protest. The cases selected were all large-scale protests that draw enough attention from the media. Second, these four cases are representative in terms of the trajectories of the development described above. They all started with negotiation and ended up with serious violent confrontation between protesters and local authorities.

Case #1: Huashui Protest (Dongyang, Zhejiang) (浙江东阳画水事件)

Since 2001, the Huaxi township government of Dongyang, Zhejiang, began constructing Zhuxi industrial park that hosted 13 chemical, dyeing, and plastics firms. Soon the villagers nearby began to complain that chemical and insecticide plants released large amounts of waste gas and water that gave off acrid smell. After several rounds of complaining without any solution from the firms and the park authority, villagers came to the town major's office on March 15, 2005, the international day for consumers' rights, and sought a solution be found by the township government. Yet they were not received by any government officials. Five days later they set up tents on the roads toward the park. The tents were guarded by elders day and night in the hope that no one dared to tear off the tents using force against these elders. The transportations to the park were blocked, so did the construction for remaining plants. Large banners were hung up everywhere with slogans such as "I will survive, I need environment!" and "we have our offspring too!" On April 6, local police issued an ultimatum-like notice that required villagers to remove tents and leave. Yet, villagers continued their protest. Before the dawn of April 10, over 100 police cars came and about 3,500 heavily equipped law enforcers began to remove the tents. To the local government's surprise, over 10,000 villagers came to confront the law enforcers and a fight broke out between police and villagers.

After the conflict, eight villagers were reported to be prosecuted and some local officials to be removed from office. The local government made some positive moves against the firms. The insecticide plants were ordered to close and all the chemical plants were ordered to relocate.

Case #2: Liuyang Protest (Liuyang, Hunan) (湖南浏阳镉污染事件)

In Zhentou township of Liuyang, Hunan, Changsha Xianghe chemical plant, starting operation in 2004, has been known for lacking environmental protection measures and releasing huge amount of toxic water into the nearby river of Liuyang. Trees died and crop outputs dropped on a massive scale. Residents living nearby soon experienced sickness with various symptoms. Hospital visits suggested excessive level of cadmium in the body of many residents. In 2006, residents began lodging complaints and petitioning but did not get any effective response. Residents also spent money to obtain the tests for the content of heavy metals in water and soil. Yet their efforts and complaints fell on deaf ears. In early 2009, two villagers died of cadmium poisoning, which caused panic among residents. On July 30, 2009, thousands of angry villagers besieged the township government building and asked for remedies and compensation. The protests finally forced the local governments at different levels to get involved and they promised to solve the problem. Next day, the owner and the key managers of the factory were arrested and the officials of the local environmental protection bureau (EPB) were investigated. The local government later declared to permanently remove all chemical plants from this region. Residents were sent to hospital to get tests and they were compensated.

Case #3: Fengwei Protest (Quanzhou, Fujian) (福建泉州蜂尾事件)

In August 2009, the one-month-old wastewater treatment plant located in the Fengwei township of Quangang district, Quanzhou city of Fujian, began to produce acrid smell. The local residents suspected that the plant was used to treat not household wastewater but the petrochemical industry wastewater. The residents had complained for years about the pollution caused by the nearby meizhouwang Chlor-alkali company, accusing it of affecting the health of the residents as well as the fishery industry, which had always been the major industry in the region. Over 20 cases of esophageal cancer were found in recent years. The wastewater treatment plant further exasperated the anger of local residents. After the complaints had failed to produce any positive response from the plant and the government, the villagers began to besiege the plant and smash some equipment. On August 31, villagers refused to allow government officials to enter the plant. A fight broke out. Some government employees were detained. The local governments later sent over thousand armed police to rescue them but it only made the situation worse. For several hours, the angry villagers were confronting the armed police, until the latter withdrew for fear of more serious conflict. In the days that followed, the plant made efforts to reduce the smell, though the government never agreed to villagers' demand to either relocate them or relocate the plant.

Case #4: Jinxi Orotest (Jinxi, Guangxi) (广西靖西县铝厂污染事件)

On July 11, 2010, over thousand villagers of Pangling village of Xinjia township in Jinxi county, Guangxi, blocked the major roads toward the county government. Complaints were written on their cloths as well as banners, such as "give us back our home and our rivers" and "clean the river, beautify Jinxi." The complaints were about the pollution caused by the Xinfa aluminum plant for many years. Without appropriate treatment after mining, the soil and the river got badly damaged. There are various explanations in media about the cause of the fight on July 11. The official media report said that the plant's construction team had a dispute with the local residents and it grew into a fight. The overseas media reports said it was because the plant workers beat the protesters outside the plant and angered the villagers. Two days later an even bigger rally among villagers blocked major roads and fought against the armed forces that came to maintain order. Many police and government cars were smashed and many people were injured. Later on, the local governments made certain compensations to villagers and also punished some officials and firm managers.

These four cases differ in time and location. They, however, resemble each other in terms of the way they developed. The following Table 7.1 compares the four cases and serves to illustrate the differences and similarities among them.

Table 7.1 Comparison of the four cases

	Huashui (Zhejiang)	Liuyang (Human)	Fengwei (Fujian)	Jinxi (Guangxi)
Cause for protest	Waste gas and water pollution	Cadmium pollution	Wastewater pollution	Aluminum pollution
Starting time	2001	2006	2009	Unclear
Early strategy of protest	Petition and open letters	Complaints to the local environmental protection bureau; letter petition; internet exposure; letter to the mayor	Besiege the plant; close down the plant; detain government employees	Petition; occasional fights with factory workers
Target	Chemical plants in the Zhuxi industrial park	Changsha Xianghe chemical plant	Fengwei Urban wastewater treatment plant; Meizhouwan Chlor-alkali company	Guangxi Xingfa Aluminum plant
Government's initial reaction	In 2003, EPB claimed that the death of rice crop was not related to water quality; in 2004, Zhejiang government claimed to close down the park, but it in fact expanded; in 2005, a chemical industrial park began to construct in Wang village of Huashui	Liuyang EPB closed down part of the operation line and fined the factory; but failed to act against the later pollution; in March 2009, EPB denied cadmium pollution	Local government promised to purchase detergents to remove the smell	No action; county government claimed that nobody opposed the construction of the plant

Table 7.1 (Continued)

	Huashui (Zhejiang)	Liuyang (Human)	Fengwei (Fujian)	Jinxi (Guangxi)
Trigger for conflict escalation	On March 15, petitioners were refused to see anyone on the major's visit day	Several villagers died	Local officials had physical fight with a local woman and that angered surrounding villagers	On July 11, the factory construct roads, which was opposed by villagers (official media); or factory workers beat protesters (overseas media)
Later strategy of protest	Set up tents and block the roads toward the park; slogans and banners	Petition to the provincial government; block the roads	Occupy the factory and detain governmental workers	Physical fight between workers and villagers; villagers later gathered to visit county government
Government reaction	Open letter; send police to remove tents by force	Deny the pollution and poisoning	Send police to crack down	Send police to crack down on the fight
Confrontation	Physical fight; multiple cases of injuries; destroy cars	Escalation; besiege township government	Villagers throwing stones; police withdraw	Physical flight but unclear about details; police crackdown on the fight
Outcome	Polluting firms were closed down and key managers were punished	Factories were closed; factory managers and protest leaders were punished	Negotiation successful; detergent was used; government pay for tests and compensate	Punish the factory managers; provide villagers clean drinking water and compensate; the factory build wastewater treatment system

3. Mass Protests and Government Reactions: A Temporal Analysis of Environmental Conflict

The above description of the selected cases suggests a clear trajectory of the environmental conflict in rural China, starting from peasants' awareness of pollution to their complaints, and ending up with a violent protest. Three stages of development can be identified in all of these cases: articulation of interest within the legal framework; massive violent protests outside the legal framework; and policy readjustment.

3.1. Articulation of Interest within the Legal Framework: The Preconflict Stage

The cases under analysis suggest that those peasants who suffered from pollution had been struggling for quite a while before the conflict took place. In the Huashui case, the peasants began to protest against the Chemical plants in 2001. Even though some sort of physical fights occurred occasionally, the massive protest took place only four years later in 2005. In Liuyang and Jinxi, the process was also quite long. Only in the Fengwei case did the process took place in months. The long process toward massive protests in most cases suggests that violent conflict over environmental pollution did not occur abruptly and unexpectedly but grievances accumulated over a period of time ultimately led to the violence.

The physical damage inflicted on peasants can promote them to articulate and defend their interests, individually or collectively. Oftentimes peasants would want to solve the problem through governmental intervention. In this sense, peasants' environmental protests are a form of political participation on a specific issue. The grassroots activists normally do not pay much attention to whether their approaches are judicial or administrative ones. Oftentimes they use multiple measures simultaneously as long as these measures are perceived to be effective (Ying 2007:17). The most common measure peasants choose is petition. Petition is a unique form of political participation in China. It is not fully compatible with the modern judicial system but it is a measure accepted by the authority within the current legal framework (Wu 2007). This form of political participation is designed for the state to co-opt social demands and reduce social unrest. It is relatively easy to access within the legal framework and therefore is widely recognized by peasants. In addition to petition, letters of complaints and circulating letters among the public are common as well. In recent years public tip-offs on internet too have become popular. These measures are not fully institutionalized but they provide a relatively peaceful channel to solve problems within the legal framework.

Governments' reaction to peasants' choice of action is important for peasants' next move. In the selected cases, local governments' initial reactions were identically passive. In the Huashi case, local EPB claimed that pollution was not responsible for the death of rice paddy. The Liuyang EPB, though once punished the polluting firm, later refused to take action against the illegal operation of the firm and denied the existence of cadmium pollution. The Jinxi case was similar. Villagers did not get any reaction from their petition. Worse yet, some of the villagers were beaten up by the factory workers and no one was held responsible for that. It is common that local governments chose inaction as their strategy against peasants' demands and hope that the complaints would fade away over time. If peasants' demands become strong, they normally use force to repress. The inaction and repression of local governments may be the result of the following reasons. First, peasants' demands were not strong enough and failed to make the local governments aware of the seriousness of the issue. Second, the environmental system is malfunctioning, preventing the local EPB to be effective (Mol and Carter 2006). Third, local governments are biased toward the polluting firms, which are important to government for revenues, local economy, and claim of officials' achievement and, as a result, local governments ignore environmental protection and peasants' interests. Fourth, local officials and managers of the polluting firms collude to defend their common interests against environment as a public good. Regardless of the reasons, local governments tend to ignore and even repress peasants' articulation of their interests. This suggests a problem with the governance at the local level—local governments are either incapable or unwilling to incorporate peasants' articulation of their interests into their policymaking and incapable of solving the existing problems.

In this sense, it is insufficient to argue that peasants' petition primarily intend to express their grievance or that their complaints resemble a family dispute (Ying 2007:10–11). These arguments fail to explain the action of the governments. Nor can they explain the escalation of conflict and the tendency for a violent ending. In response to the complaints and demands of peasants, the inaction and denial of local governments only further reduce the space for peasants to express their opinions. As a result, desperate peasants either continue to suffer the damage from pollution or make their action even more noticeable. Once peasants' grievances were turned into hatred and they no longer trusted the governments, massive violent protests may result at any time.

3.2. Massive Protest Outside the Legal Framework: Conflict Stage

When their complaints through legal channels were ignored or their demands refused to be addressed by the firms and local authority, peasants would resort to more effective measures and their targets often turned from firms to local government.

In all cases, peasants' protests escalated thanks to the inaction of firms and local governments. In Huashui, Liuyang, and Quanzhou cases, villagers blocked roads to the factories or occupied the factories. In the Huashui case particularly, villagers' confrontation was more symbolic. They set up tents on the roads to the factories and asked village elders to stay in the tents. Such theatrical maneuvers were meant to put pressure on local authority. And yet, the local government sent police and law enforcers to remove the tents in the early morning. It was such a response from the local authority that turned the theatrical but otherwise peaceful protest into a violent confrontation.

In the Fengwei case, villagers' action involved certain amount of violence from the beginning. Because villagers in this case lacked self-organization and discipline, any conflict could easily turn into violence. In this case, the trigger for a possible massive fight was that a local official had some physical conflict with a female villager and was immediately taken seriously by other villagers. Once the official was detained by villagers, local police pressured villagers and tried to use force to rescue the official. What is noticeable in this case, though, is that the local government, realizing the potential fight that might follow, immediately withdrew the police force and thus avoided the escalation. This contrast between the two cases suggests that local governments' reaction determined the possibility and the scale of massive violent confrontation.

In addition, villagers' distrust of the government authority was another important factor responsible for the escalation. In the Liuyang case, the local EPB until 2009 had constantly refused to admit the violation by Xianghe chemical factory and refused to acknowledge the existence of cadmium pollution. After the peasants began to use road-blocking to force the local government to take measures to punish the managers and compensate the victims, peasants had already had no trust at all in the local government. Evan after the government announced permanent closure of the factory, the anger among villagers did not subside (Yang 2006). In the Fengwei case, after the government promised to use chemicals to reduce odor in the wastewater treatment plant, villagers believed this to be a trick used by the government to destroy the evidence (Zheng 2009). In this environment of distrust, peasants preferred violent measures in the hope of drawing attention of and obtaining intervention by higher-level governments for a once-for-all solution. Unfortunately, such violent protests often propelled local governments to react—and also gave them legitimacy—using radical approaches such as deploying police force to put it under control, which is a zero-sum game. In brief, the serious deficiency of the trust in local authority has become an important impetus for violent conflicts between governments and citizens in rural areas.

3.3. Adjustment of Interests: The Post-conflict Stage

After the violent confrontation, peasants' demands would normally enter into the policymaking agenda of local governments and be partially considered. The interests among peasants, firms, and local government would be readjusted at a certain level. That means the illegal approach normally would give peasants certain positive outcome, though in the four cases the results varied considerably depending on circumstances. In all cases except Fengwei, firm managers were punished in certain ways. In Huashui and Liuyang, polluting factories were closed and some local officials were punished for either failing to curb pollution or for mishandling the protest. Punishing certain local officials could pacify peasants. Yet their distrust remained strong afterwards.

The above description suggests the following pattern of rural environmental protests: At first, peasants' protests begin by peaceful and legal means. It often takes quite a while before the peaceful protests turn into violent, illegal ones. Second, the escalation of conflict from legal to illegal was often the result of the fact that peasants' demands were ignored or not met at a reasonable level and their complaints were not heard through institutionalized channels. Third, the initial target of peasant protests was in most cases the polluting firms, and yet local governments' mishandling of protests eventually turned themselves into the target. Fourth, the distrust among peasants toward local governments played an important role in the escalation.

4. Local Governance Deficiency: A Structural Analysis

Peasants' protests are determined by a verity of factors including the magnitude of pollution and the organizational structure and resources available for villagers such as social norms within the rural community. At the same time, equally important is the interplay between local governments and peasants and the strategies of each side. If peasants' interests were protected through legal channels, illegal means such as violent confrontation can be avoided. The above description demonstrates that massive protests and their escalation into violent confrontation were out of desperation of peasants due to the inaction of local governments. The peasant protests and their escalation illustrate the governance deficiency facing rural local governments in three aspects: conflict of interests, governability crisis, crisis of trust. The three aspects interact with one another and determine the behavior of local governments and the possible escalation of peasant protests from legal to illegal ones.

4.1. Conflict of Interests

Local governments face a large number of interests groups. Among these, firms often have strong influence and oftentimes they form alliance with governments against other interests. After the traditional model of combination of government function and enterprise managements was abandoned since reforms began, the line between firms and local governments became somewhat clearer. However, the current political system produces the so-called pressure-dominated system, which has tremendous consequences on local governance (Rong and Cui 1998). In this system, the central government maintains control over local governments by imposing various economic growth targets in the personnel evaluation system. This system works well in motivating local officials to improve GDP growth and has become an important dynamic for China's high growth rate. The downside of it, however, is local governments' ignorance of environmental impact of the GDP-oriented development model. In addition, the decentralization process after reforms has given local governments power to manage economic development, which further motivates them to pursue short-term goals of growth at the expenses of environment (Schwartz 2004). As a result, the combination of evaluation pressures from the central government and the increased economic autonomy they have enjoyed has pushed the local governments to adopt an opportunistic strategy in dealing with local interests. More specifically, they support firms at the expense of environment and the wellbeing of local citizens. A third factor aggravating this situation is the fiscal pressure on local governments. To maintain the function of an enormously large local bureaucracy with excessive personnel and benefits, local governments are under pressure to expand local revenue, and the best way to do so is to turn to local enterprises for help. Oftentimes local governments make efforts to cultivate and support local enterprises for taxes and revenues (Zhang 2006). Since local governments have to compete to carter investment, the pressure creates a "race to the bottom" effect in which local governments compete to reduce the standards for environmental protection in order to attract firms. Lastly, the lack of monitoring from both above and below gives local officials tremendous opportunities to engage in rent-seeking activities, which further reduce the incentives for them to go against the polluting firms.

These factors, working together, make it inevitable for local authority and economic interests to collude and even institutionalize their alliance. The traditional model of government–firm combination has transformed into a new type of government–business collaboration that dominates development in most localities (Zhang 2006). This system is fundamentally responsible for motivating local governments to ignore or repress peasants' environmental complaints and preventing local governments from making positive contribution in local environmental governance.

In the Liuyang case, for example, the deputy township governor of Gejia township had personal shares in Xianghe chemical plant and regularly took bribes from the firm managers. In Jinxi case, the collusion of interests between the government and firm was more striking. The CEO of the polluting company also serves as the deputy party secretary of Jinxi country, in charge of industries. The cozy relationship between the government and firms in these cases was an obvious cause for the conflict of interests in local governance. Under these circumstances, peasants' environmental demands were difficult to get addressed using legal means. As a response, they were forced to adopt radical, illegal means and change their target from firms to local authority in order to break the government–business collusion. In other words, peasants' massive protests were in most cases not a "government–peasant" conflict in the first place. They originated from a "firm–peasant" conflict. Against this government–business collusion, a feasible way for peasants to defend their interests is to get noticed from upper-level authorities and media in order to break the entrenched structure.

4.2. Governability Crisis

The cases analyzed suggest that inability of local governments to handle peasant protests was very critical in stimulating and even producing environmental conflicts. Governability refers to the ability of the government to implement state goals, particularly against the potential and actual opposition of strong social interests or under the unusual circumstances of economic crisis (Skocpol 1985). In the context of environmental protests in rural China, the lack of governability refers to two things. First, local governments were incapable in integrating the various interests, particularly the interests of peasants, into its policymaking. Second, local governments were incapable in fighting against the interests of firms in enforcing environmental regulations.

As the cases have illustrated, local governments often failed to incorporate peasants' interests into their policies. And they were incapable of identifying or ignorant of the damage conflicted on peasants' health and properties. These were the factors that incited peasants to resort to illegal means to defend their interests. In other words, the fact that local governments were incapable of balancing the social interests in a timely manner due to various reasons reflects a problem in the current system, that is, the lack of responsiveness of local authority facing powerful articulation of interests. The current system lacks effective channels for peasants to express their interests and participate in policymaking. Petition becomes the most common yet ineffective channel. The weak organizing ability among peasants, weakened grassroots governmental bodies such as village governments and party branches, and peasants' insufficient access to media, working together, limit the channels available for peasants to articulate and integrate their interests. Under such circumstances, as Almond and Powell argue, where social elements lack an organized collective or representative to articulate their interests, either an accident or the availability of a capable leader could easily intensify the grievance accumulated among peasants and provoke a conflict that is often unpredictable and difficult to control (Almond and Powell 1978).

Local governments' lack of governability is also reflected in their inability to control the polluting firms. In the cases analyzed, most plants in operation had no basic environmental protection measures or never carried out environmental assessment. Local EPBs lacked resources in every aspect including personnel, capital, skills, and technology, which are needed to meet the requirements for environmental protection. Instead, they often turn environmental enforcement into a profitable business by charging fees on firms so they can pollute legally. This in turn renders local EPB officials vulnerable to powerful business interests.

4.3. Trust Crisis

The distrust among peasants in local authority often becomes an important trigger for environmental conflict and it often results from the interaction between the two parties over a long period of time. When peasants' interests have been ignored persistently or even repressed by the local government, distrust grows quickly among them. Distrust then makes peasants to desire other means to draw attention from higher-level authorities. Thus their protests are often highly confrontational and theatrical for the purpose of drawing attention.

During the process of protests, peasants consciously differentiate the insiders and outsiders and form their identity based on their residential areas and the common damage inflicted on them. They often identify firms and local governments as outsiders that have conflict with their interests. As conflict continues and the self-identity becomes clear, the truth becomes less important. Any action of local governments can potentially be interpreted as tricks. Peasants no longer trust whatever local governments say and do but instead prefer using other approaches to draw attention of media and high-level authorities. As the peasant wisdom goes, "making a small trouble, you got a small-scale solution; making a big trouble, you got a big solution [*xiaoniao xiao jiejue; danao da jiejue*]." The violent and illegal—therefore dangerous—measures peasants often resort to are radical but not irrational. Fundamentally, they choose to do so because they no longer trust local governments but instead expect the higher-level authorities to intervene and offer them help.

The three aspects of local governance deficiency—conflict of interests, governability crisis, and trust crisis—are intertwined and they interact among themselves. However, the conflict of interests in local governments is a major cause not only of environmental damage in rural areas but also of the repression of peasants' voicing of their interests. It weakens local governments' governability and damages the trust in them among peasants, even though the governability crisis itself can have direct impact on peasants' trust.

5. Conclusion

Using four cases of environmental protests, this study has demonstrated that local governments' governance deficiency is fundamentally responsible for the escalation of environmental protests from a peaceful and legal conflict to a massive-scale, violent, and illegal confrontation. Environmental participation among rural residents is not always characterized by violence in the first place. Instead, peasants in most cases would use means that are legal and acceptable to polluting firms and local authorities. But when their demands are ignored or repressed by the firms and the local authority, they might use illegal and violent approaches under such circumstances and switch their target from firms to local governments. The massive violent confrontation is fundamentally a passive approach for peasants in dealing with environmental damage inflicted on them. Nevertheless, under the current system it is an effective approach for peasants to force local governments to correct their wrongdoing or at least restrain their behavior to a certain degree.

Changing the governance deficiency among local governments is a critical step in reducing and solving rural environmental conflict. In doing so, local governments need fundamental reform to change from an "entrepreneurial government" to a "service government" and break away with business interests. The central government is critical in this process by changing the evaluation system, strengthening the environmental regulatory system, and introducing stricter environmental standards for local governments to follow. Meanwhile it is essential to introduce the formal channels that allow the public and societal organizations to monitor local governments. Peasants should be allowed to participate in environmental policymaking whenever their interests are involved. The last but not the least, rebuilding the peasants' trust in their local authority is a painful but critical and urgent task for both the governments and the society so as to restore the legitimacy of the government at the local level.

BIBLIOGRAPHY

Almond, Gabriel and Bingham Powell. 1978. *Comparative Politics: System, Process, and Policy* (2nd edition). Boston: Little, Brown and Company.
Lang, Youxing. 2005. "Deliberative Democracy and Public Participation in Environmental Governance in China: An Empirical Research from Zhejiang's Two Cases [*shangyixing mingzhu yu gongzhong canyu yuanjing zhili: yi Zhejiang nongaming kangyi huanjing wuran shijian weili*]." in Proceedings of the International Conference on "Development of Deliberative Democracy [*xieshang mingzhu de fazhan*]." Beijing: Chinese Social Science Press.
Mol, Arthur and Neil Carter. 2006. "China's Environmental Governance in Transition." *Environmental Politics* 15(2):149–170.
O'Brien, Kevin. 2009. "Rural Protest." *Journal of Democracy* 20(3) (July):25–28.
O'Brien, Kevin and Lianjiang Li. 2006. *Rightful Resistance in Rural China*. New York: Cambridge University Press.
Rong, Jingben and And Zhiyuan Cui. 1998. *Transition from the Pressure-Dominated to Democratic Cooperative System—Political Reforms at Township and County Levels* [*cong yalixing tizhi dao mingzhu hezuo tizhi de zhuanbian—xianxiang liangji zhengzhi tizhi gaige*]. Beiing: Central Compilation and Translation Press [*zhongyang bianyiju*].
Schwartz, Jonathan. 2004. "Environmental NGOs in China: Roles and Limits." *Pacific Affairs* 77(1):28–49.
Skocpol, Theda. 1985. "Bringing the State Back In: Strategies of Analysis in Current Research." In Peter Evans, Dietrich Rueschemeyer and Theda Skocpol (eds.) *Bringing the State Back In*. New York: Cambridge University Press.
Wu, Yi. 2007. "The Structural Net of Power-Interest and the Predicament of Peasants' Collective Interests Articulation [*quanli-liyi de jiegou zhiwang yu nongaming qunti liyi de biaoda kunjing—dui yiqi shichang jiufen de anli fengxi*]." *Sociology Studies* [*shehuixue yanjiu*] 5:21–45.
Yang, Yongxing. 2006. "Unhappy about Aluminum Pollution, Massive Protest broke in Guangxi Zhuang Minority Region [*buman luchange wuran huanjing, Guangxi Zhuangzhu fangsheng quntixing shijian*]." *Singapore United Morning Post*, October 16.
Ying, Xing. 2007. "Grassroots Mobilization and the Mechanisms of Peasants' Interests Articulation [*caogeng dongyuan yu nongaming liyi de biaoda jizhi*]." *Sociology Studies* [*shehuixue yanjiu*] 2:1–23.
Yu, Jianrong. 2004. "An Analytical Framework of Current Peasant Rights Activities [*dangqian nongming weiquan huodong de yige jishi kuangjia*]." *Sociology Studies* [*shehuixue yanjiu*] 2:49–55.
Zhang, Yulin. 2006. "The Developmental Mechanism of Government-Business Combination and China's Rural Environmental Conflict [zhengjing yitihua kaifajizhi yu zhongguo nongcun de huanjing chongtu—ye zhejiangsheng de sanqi quntixing shijian wei zhongxing]." *Exploration and Discussion* [Tansuo yu zhengming] 5:26–28.
Zheng, Dongyang. 2009. "Violent Anti-Pollution Protest in a PX Port [*yizuo shihua gangkou de baoli kangwu shijian*]." *Phoenix Weekly* (HK) V 28:46–49.

CHAPTER 8

ENVIRONMENTAL NGOS AND
PARTICIPATIVE GOVERNANCE: THE
CASE OF THE PM2.5 INCIDENT

Zhen Lin and Yin Guan

1. INTRODUCTION

Thanks to the constant efforts of environmental NGOs and environmentalists during the winter of 2011, "PM2.5" has changed from a mysterious professional jargon into a panic-driven household name and drawn a great attention from the society. This change occurred within just three months that the term "PM2.5" first came into public view in November 2011 when the State Council gave its consent to release the newly revised *Ambient Air Quality Standard*. This new standard includes PM2.5 concentration measuring index. The following three months witnessed not only how China's green NGOs and the public inspired by the pollution jointly participated in environmental protection and finally turned the environmental crisis into a collective policymaking event, but also how the relations among the Chinese government, environment authorities, and green NGOs evolved and improved.

It is often believed that the Chinese government leads an absolutely authoritarian position in the environmental governance, and that it is hard for green NGOs to play a substantial role in it. However, with the serious environmental problems in China and the growing attention to ecological civilization [*shengtai wenming*] by the ruling party and the government, green NGOs have emerged to play an increasingly important role in China's environmental protection. Green NGOs differ from the NGOs in other fields. Their main activities and mission are to protect and improve environment. They are also different in their highly professional performance, strong relevance to the wellbeing of the entire society, and their rapid growth. This study provides an opportunity to understand how green NGOs have played an increasingly important role in China's environmental governance. In what follows, we will analyze the model of participation of Chinese green NGOs and the public in the environment protection and investigate green NGOs' strategies to push the government forward to change attitudes toward environmental protection and public participation as well as making new policies regarding PM2.5.

2. Environmental Problems and PM2.5

To many Chinese people, PM2.5—inhalable suspended particles amidst atmosphere with a diameter less than or equal to 2.5 micrometers, or less than one-twentieth of that of hair—is a brand new concept. Scientists use "PM2.5" to indicate the content of particulate matter in every square meter of air. The higher the content, the severer the pollution. Because a great amount of toxic and harmful substances in PM2.5 stays long in air and can travel long distance and penetrate into the human body through lower respiratory tract, PM2.5 is quite detrimental to human health.

The air quality in Beijing has been deteriorating over the years but the situation became very severe in the winter of 2011. While the air quality monitored by the US Embassy in Beijing turned out to be "crazy bad," it was only "slightly polluted" according to the official data released by the Ministry of Environmental Protection of PRC, which used the PM 10 standard by that time, thus giving rise to the striking disparities. When Beijing was shrouded in thick fog and haze for days in a row, people had to wear masks whenever they were in open space, and became restless over the terrible air quality, poor visibility, and soaring incidences of respiratory diseases. This particular environmental problem finally grew into a public issue.

The PM2.5's earliest use in China can be traced to the Pearl River Delta Region. Simultaneously with the rapid industrialization process in the recent years, PM2.5 has been continually accumulating through the emission of manufacturing waste, automobile exhaust, and construction dusts. Air quality problem has long been a big headache in Northern China. Around the Spring Festival of 2013, thick fog reined more than one million square kilometers of the territory across the nation for a long period. Beijing was particularly hit hard and as a result gained two humiliating nicknames as the Capital of Fog and the Capital of Toxin.

Though grave in nature, the air pollution caused by PM2.5 has long failed to draw sufficient attention of the nation. The problem did not win public reorganization until US Embassy in Beijing started to measure and publicize their records of PM2.5 since 2008. Among the first batch of environmentalists and Microblog celebrities who were concerned about the problem, Ma Jun, Director of Institute of Public & Environmental Affairs; Feng Yongfeng, Founder of Nature University of Green Beagle Environment Institute; Wang Yongchen, convener of Green Earth Volunteers; Pan Shiyi,

Chairman of SOHO China; Zheng Yuanjie, famous writer—all quoted the PM2.5 data released by US Embassy and compared it with those released by the Ministry of Environmental Protection. As an ultimate provocation following a string of problems affecting the normal life of common people such as thick fog cover, frequent breakout of respiratory diseases with symptoms of cough and difficulty in breath, and poor visibility, the PM2.5 incident detonated public indignation.

As the air pollution crisis aggravated, many people advocated including PM2.5 into the governmental air quality report. This public demand quickly transformed the problem into a public policy debate, and some government authorities and social organizations were expected to incorporate the debate into legislation agendas and work out well-targeted policies as solutions. Positive signals were released from the Ministry of Environment Protection in November 2011 in the wake of the PM2.5 incident. Efforts were underway to explore ways to modify the outmoded air quality measurement system.

The soon revised *Ambient Air Quality Standard* added additional measuring indexes such as the concentration of PM2.5 and ozone (O3) over a period of eight hours into the new measurement. On February 29, 2012, it was ratified by the State Council Executive Conference chaired by Prime Minister Wen Jiabao. Meanwhile, PM2.5 was also written into national standards and brought into compulsory supervision codes by relevant provincial and municipal authorities. Also in the conference, a detailed timetable was laid down for cities across the country to start reporting the concentration of PM2.5 and ozone (O3) over a period of eight hours. Key areas including Beijing, Tianjian, Hebei, the Yangtze River Delta Region, the Pearl River Delta Region, as well as municipalities directly under the central government and provincial cities were demanded to kick off such enhanced supervision starting 2012; 113 additional key and national environmental protection model cities from 2013; all prefecture-level cities from 2015. Up to now, Beijing has put PM2.5 under close monitoring and made it known to the public.

3. Understanding Green NGOs

In recent decades, "green" has evolved into a word that is popularly used in fields related to environmental protection. Meanwhile, the society has also witnessed the birth of many green-related social and political norms like green party, green enterprise, green consumption, just to name a few. Green represents a life pattern in which people enjoy harmonious coexistence with nature. Compared with traditional terms like environmental protection, the term of "green" can do a better job in demonstrating the change of lifestyle in today's society. The term of environmental protection for its traditional meaning has failed to cover the contents of all activities conducted by NGOs in this filed.

Nongovernmental environmental protection organizations are mentioned in various ways in literatures, such as civil environmental protection organizations, environment groups, environment NGOs, green NGOs, etc. The research on them is conducted from various angles and for different purposes. In this chapter, all nongovernmental organizations dedicated to promoting China's eco-civilization undertakings are termed green NGOs.

In addition to the common characters of NGOs in general, such as being independent, promoting common goals, voluntary participation, formal organization, legitimate establishment (Salamon and Sokolowski 1999), green NGOs have their unique features such as high professionalism, strong relevance, and rapid growth. Since green NGOs' activities are mainly centered on the mission of protecting and improving the eco-environment, which people rely on for survival and prosperity, green NGOs are in fact a self-conscious endeavor collectively contributed by people who have a considerable level of green awareness, and demonstrate a great sense of responsibility, participation, and sacrifice for green undertakings (Wang 2008). In addition to functioning as the green partner of government, green NGOs are also an independent component of social autonomy, operating under new governance ideas.

4. Green NGOs' Participation in PM2.5 Governance: An Analysis

The PM2.5 incident finally came to an end after the new indexes, including PM2.5, were incorporated into *Ambient Air Quality Standard*. This outcome not only demonstrates the positive attitude and active response of the Chinese government in dealing with environmental problems, but also shows the enhanced role played by the civil society with green NGOs. Green NGOs serve as the prominent representative of the public in influencing government's policymaking and participating in environmental governance along with authorities. It marks a solid step for the public to participate in environmental governance and improved mutual trust between the government and green NGOs through increased interaction.

However, in retrospect of the whole event, from the breakout of the incident to the revision of Ambient Air Quality Standard, and then to the joint efforts made by all parties to address air pollution prevention and treatment, the process of green NGOs' participation in environmental governance and the cooperation between green NGOs and the government did not always progress in a smooth manner. A description of the process will help reveal the complicated nature of China's evolving environmental governance and participation.

4.1. Change Official Understanding of Pollution and Environmental Participation

At the very beginning, government authorities and green NGOs have quite different attitudes toward the unexpected PM2.5 incident. With great disparities in perception, both sides had gone a long way until finally reaching a consensus on jointly governing this public affair, as marked by the release of the revised standard.

Initially, environmental authorities remained uninterested toward the measurement data that were released by the US Embassy in Beijing and promoted by local environmentalists. Some officials even believed it was a "deliberate" political confrontation under the disguise of environmental controversy. While the Chinese government complained loudly about the practice of the US side, the spokesman from the US State Department insisted that the US Embassy was collecting data about the quality of air in certain Chinese cities and that had nothing to do with the so-called intervention in China's internal affairs. Instead, it was only for the purpose of providing information to US citizens to understand the situation in China. For a while this environmental problem escalated into a political issue with growing tendencies of confrontation from both sides.

Amid the controversy, many green NGOs and environmentalists stood out to express their opinions: instead of fighting endlessly with the US side for the right to release air quality data, it was wiser for China's environmental protection authorities to seize the opportunity to promote China's environmental protection by adopting PM2.5, which was used by the US Embassy, into China's air quality evaluation system so as to promote public awareness of environmental protection and pave the way for clean industry investment. What's more, green NGOs quickly seized the opportunity to promote environmental protection undertakings. For example, campaigns such as "Let Me Measure Air Quality for My Motherland" launched by the Green Beagle Environment Institute and "Blue Sky Plan" initiated by the Institute of Public and Environmental Affairs all drew great attention and wide recognition from the society and also greatly boosted the profile of these organizations. Meanwhile, as part of efforts to exert influence on policymaking, green NGOs allied with each other to call on the government to incorporate PM2.5 into China's air quality evaluation system and put energy-saving and emission-reducing measures well in place.

Confronted by waves of public uproar with sarcastic rhetoric, like "doomed victims of the noxious wind and air," officials in the government and environmental protection authorities finally changed their way of looking at the issue. With a changed attitude, officials became much more open and candid in their public remarks about PM2.5. "To drink polluted water while driving BMW—what an ironic modern scene it is," remarked Mr. Zhou Shengxian, then Minister of the Ministry of Environmental Protection. Finally, the timetable to start putting PM2.5 under supervision was rolled out after many rounds of meetings and consultations. Following the central government's steps, local governments became aware that "it is better to correct than to hide problems." Shanghai proposed a policy to issue a PM2.5 report daily. Nanjing disclosed its PM2.5 data for the last four years.

4.2. Push Public Affairs into Policymaking Agenda

In addition to launching publicity campaigns for PM2.5, green NGOs, such as Institute of Public & Environmental Affairs, Green Beagle Environment Institute, and Friends of Green Environment, also exerted their every influence to change the policymaking style of the government. Equipped with air quality monitoring devices, NGOs and many residents set out to measure PM2.5 on their own and publicized it via the Internet. Civil self-salvation campaigns spread over the nation. a nationwide campaign, "Let me Measure Air Quality for My Motherland," organized by Green Beagle Environment Institute among other green NGOs, grew very fast. Though some environmental experts and officials remained somewhat skeptical about the scientific accuracy of the data measured by green NGOs, such campaigns themselves were, without doubt, an ideal way for NGOs to have their voices widely heard.

These large-scale campaigns spread rapidly via the Internet and the emerging social media, drawing great attention from environmental protection authorities as well. In addition, green NGOs teamed up with environmentalists and meteorological experts to advocate revising *Ambient Air Quality Standard*. As a result, the revision process was hastened and PM2.5 was also incorporated into the standard to form a more complete and rigorous measuring system for the future supervision. In China, the agenda of public affairs are usually proposed by officials before they go into policymaking agenda for deliberation and discussion. This time, strikingly enough, green NGOs took the lead in pushing public affairs into a policymaking agenda, fully demonstrating their influence.

Without the efforts of green NGOs in publicizing PM2.5-related knowledge, it was unlikely for the public to gain a quick understanding of the damages caused by PM2.5 on environment and humans. Thanks to the newly available social media, green NGOs are able to exert their influence in a fast and effective way. Friends of Nature, for example, with over 30,000 followers in Sina weibo (or Microblog, the Chinese equivalence of twitter), published 30-strong postings regarding PM2.5 in the platform from October 1, 2011 to May 13, 2012. Ma Jun, Director of Institute of Public & Environment, with 17,000 followers in Sina weibo, published 210-plus postings regarding PM2.5. Using various publicity channels available on Internet, social networking tools, and traditional media, green NGOs successfully made PM2.5, a highly professional jargon, an easy access to the public at a fast pace. It is safe to say that no large-scale scientific publicity endeavor in the nation ever was as triumphal as the PM2.5 incident in terms of far-reaching influence, wide scope of coverage, and attention-grasping effect.

Government authorities also joined the efforts for disseminating information and knowledge about environment protection through the channels employed by green NGOs as well as their own. Many government branches opened their own weibo accounts to facilitate their communication with the public. The effect appears to be positive. During the 2013 Spring Festival,

THE PM2.5 INCIDENT   181

for instance, the initiatives of "keeping away from firecrackers, keeping away from fog and haze" launched by some local governments worked well to reduce accidents and haze. In Beijing alone, use of firecrackers decreased by 40 percent.

4.3. Improve the Quality of Environmental Governance

Green NGOs not only accelerated the revision and implementation of environmental policies but also promoted governments to improve their governance competency in addressing environmental problems. This has four aspects.

The first is to improve governability. Following the PM2.5 campaign in the winter of 2011, environmental protection authorities and related industries at both national and local levels made new policies and measures in order to cut down emission, save energy, and improve air quality. Take Beijing for instance. Much more rigorous measures were adopted to bring down the consumption of coal, eliminate the worn-out vehicles, restructure industries, and phase out high-pollution firms. On January 6, 2012, for the first time, Beijing published its monitoring data on PM2.5. On February 17, the city government issued *Opinions on Implementation of the Document of the State Council on Key Missions for Consolidating Environmental Protection* and set an "environment priority" principle: all indexes related to environmental quality, including total amount of pollutants and PM2.5 concentration, will be put into the cadre valuation system at all levels. To date, 35 PM2.5 monitoring stations are evenly located among its urban districts and neighboring counties, giving birth to a multidimensional air quality monitoring network. Meanwhile, air quality and pollution alarming and reporting mechanisms were improved as well, with *Beijing City Emergency Plan for Heavily Polluted Days* released and put into effect. Firms and buses are required to stop operating on highly polluted days.

From early 2013, the Ministry of Environmental Protection, Legislative Affairs Office of the State Council, and the People's Congress had reached a consensus on revising the air pollution prevention and control law. The revised law not only raised the penalty cap on firms violating waste discharge regulation to one million *yuan* but also lifted the penalty cap for air pollution accidents. In addition, the revision and legislation processes for the Environmental Protection Law—the so-called fundamental environment law—as well as Ordinances on Prevention and Treatment of Vehicle Pollution, are expected to be accelerated. New measures such as "waste emission license" and "joint defense and control on air pollution" are expected to take effect within the year.

The second aspect of the improved governance is to make environmental information transparent. Green NGOs across the nation worked together to spread PM2.5 information and organize PM2.5 measurement activities. As a result, information about environment and air quality was exchanged across the country. Friend of Nature and other green NGOs started to measure and release their own PM2.5 data in various locations, providing valuable information for local residents. For example, the Institute of Public and Environment collected the PM2.5 data from authorities and private parties in the nation and released them to the public through its own website. It also conducted its own in-depth research and analysis over these data and compared their results with those by academic researchers.

Another evidence of green NGOs' impact on the transparency of environmental information is their push for official publication of PM2.5 data. China's meteorological authorities had been monitoring PM2.5 for years but never published the result to the public. In January 2010, China Meteorological Administration imposed a cap on the daily maximum value of PM2.5 in its publication titled the *Observation and Forecast Levels of Haze*. At the end of 2010, the Ministry of Environment Protection also solicited opinions from the public about whether to put PM2.5 into measurement scope. However, the proposal was abandoned due to different opinions. As a result, PM2.5 measurement data were only made accessible to researchers for scientific research or experimental purpose but kept secret from the public. Thanks to green NGOs' efforts during and after the 2011 PM2.5 accident, the data now are required to be available to the public.

The third aspect of improved governance is public participation in environment issues. After the accident, everybody seems to be able to talk about PM2.5 and its effect. Being familiar with PM2.5 and concerned about its impact on their personal health, the public now has shown growing interests and concerns about environmental problems. Moreover, their knowledge and concerns make them interested in environmental governance as well and willing to address this problem through participation. As a result, both government and the public alike now are aware of the rights of the public to access information, environmental protection authorities are urged to provide better public services, and the public becomes more inspired and interested in participating in environmental protection.

As evidence, the popular Midi Music Festival chose PM2.5 as its theme in 2012, in a bid to call greater attention to the thick fog and haze weather. As the traditional saying goes, "Amid firecracker sound coming the New Year." However, the time-honored custom of setting firecrackers in celebrating the arrival of New Year partly contributed to the new record of PM2.5 level. In Beijing alone, over six hundred million Chinese *yuan* worth of fireworks were set off every year. The longstanding custom finally came into limelight as a source of serious environmental pollution. Thanks the efforts of green NGOs, local governments took measures to encourage their citizens to use a lesser amount of fireworks in the 2013 spring festival and the result, as mentioned earlier, was impressive.

The fourth aspect of the improved environmental governance is the improved relationship between governments and green NGOs. Though environmental and atmospheric experts at home and abroad have been advocating the inclusion of PM2.5 into air quality evaluation standards, few achievements were made until 2012. Green NGOs have played an active role in recent years in providing data and valuable suggestions to the environmental authorities for better regulation and policies. As a result, these information and proposals have been brought into government discussion and played an active role for the formation of new policies and regulation.

Media often praised the success of PM2.5 accident as a "defense war" won by the public will against government bureaucracy. Yet, the case analyzed suggests that environment problems cannot be single-handedly addressed by green NGOs and the public but require governments at all levels to intervene. The mutual understanding between governments and green NGOs determined the outcome of this campaign and in fact has deepened along with their confrontation and, later, cooperation. As the gap of perceptions of two sides gradually became narrow, their relationship also changed from confrontation to mutual understanding. It was the joint efforts taken by both sides that eventually solved the problem.

In sum, green NGOs and environmentalists played a critical role in producing a positive outcome out of the PM2.5 incident. Compared with other environmental accidents occurring in recent years in China such as ConocoPhillips Oil Spill and Xiamen and Dalian's protests against the PX projects, the PM2.5 incident yielded more positive results through a so-called democratic environmental protection mechanism in which green NGOs were involved in policy change. The incident marked an important step in improving China's environmental governance. Though occurring abruptly and causing great concerns and even embarrassment for the government, it eventually helped the government to change attitudes and style of governance, drew the society's great attention to environmental protection, disseminated the knowledge of environmental protection, and helped the public to become more aware of the importance of environmental quality and their environment-related rights.

5. Mechanism of Green NGO Participation in Environmental Governance

To further understand the implication of the PM2.5 accident, figure 8.1 presents a model that helps illustrate the mechanism underpinning green NGOs' participation in environmental governance. The model is based on the experience of the PM2.5 incident but the mechanism will be suggestive of our understanding of other cases related to environmental protection in China.

In this model, green NGOs kick off the process of policy revision once they become aware of an environmental issue such as PM2.5. They will put forward the issue to the authority and/or take action to address the issue. In this process, green NGOs may cooperate with firms concerned about environment to influence public policy using firms' professional capability and other resources, so as to draw attention of the public, government, and other firms in the hope of solving the problem jointly. In this process green NGOs play the role of a coordinator and a motivator. If the problem is delayed due

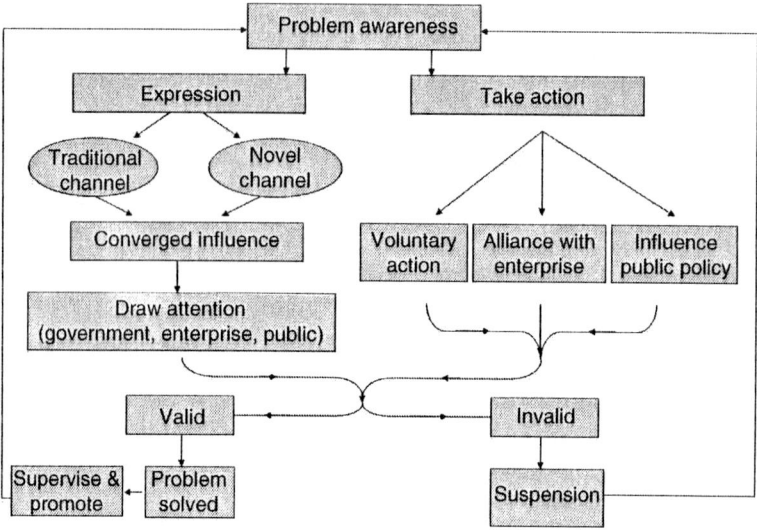

Figure 8.1 Model of Green NGOs in Environmental Participation

to lack of attention or solution from relevant parties, green NGOs resort to other means and actions to expose the problem. The process will repeat until the problem is solved.

Green NGOs disclose the problem through two channels. The first one is the traditional official channel in which written letters are submitted to the authorities concerned to required information or complaints are filed to these authorities. The second channel (the novel one)helps publicize the problem via mainstream media such as televisions, radios, newspapers, as well as the emerging media such as blogs, emails, videos, and social networking media such as microblogs. The novel channels are becoming increasingly influential and effective since they are low-cost and effective in magnifying the influence of damages caused by the environmental problems, and therefore more frequently adopted by green NGOs. In addition, major media outlets are eager to show their social conscience and social responsibility and willing to echo green NGOs' aspiration. This gives green NGOs opportunities to make the most of these media outlets' popularity to make environmental problems widely known and, meanwhile, also boost their own popularity and influence.

The environment issue promoted by green NGOs eventually will grow into a social issue and draw attention of both the public and governments. In addition to green NGOs, other stakeholders may get involved and interact with each other and form a network of governance, which eventually helps the environment problem to be solved. In this process, NGOs are most successful in cooperating with firms and other social and economic groups in order to gain funding and technological support for obtaining information such as aerial detection, sewage treatment, garbage disposal, etc., which the green NGOs may find difficult to do on their own. However, it is important for green NGOs to conduct their own research, particularly the work they are capable of doing, such as investigation, survey, and data analysis, in order to provide reference information to policymakers and related authorities.

Two outcomes may follow. Green NGOs' efforts may bring positive effects and achieve expected results by persuading governments to change policies and laws. If such changes are made, it is likely that green NGOs may be allowed to continue their influence in monitoring. If their voice and actions fail to draw enough attention from the government, green NGOs have to continue to search for other opportunities to solve the problem in the future. Their impact, however, will stay for future action.

6. Conclusion and Suggestions for Improving Green NGOs' Environment Participation

The case study of the PM2.5 incident of 2011 has important implications for understanding how to make environmental governance in China better. The main lesson we can learn from this study is that green NGOs can play a critical role in improving environmental governance. Yet, the successful cases in which green NGOs made positive contributions as they did in the 2011 PM2.5 campaign have been rare. The political barriers and social difficulties (see relevant chapters in this volume) obviously are responsible for green NGOs' limited influence. Yet it is also critical to understand the limitation of green NGOs in today's China in order to boost their influence in the future. A few issues can be identified to understand the weakness of green NGOs in today's environmental governance, and efforts must be made accordingly to make green NGOs' role in China's environmental governance indispensable.

First, green NGOs must focus more on the importance of network governance participation strategy. Network governance refers to a unique governance model where governments, citizens, enterprises, NGOs, and other social organizations work together through cooperation and interaction so as to safeguard public interests by deploying the best resources for the sound management of public affairs (Chen 2003). For environmental authorities, they should make more channels and opportunities available for green NGOs. For example, to outsource environmental evaluation projects to some third-party green NGOs with high professional performance; or to work with green NGOs to conduct environmental and eco-civilization publicity and education programs. On the part of green NGOs, they should be bothered by the dilemma between "obedience" and "resistance" in dealing with governments but instead utilize whatever resources are available to them in order to make positive contribution to the changing attitudes and behavior of the government so as to solve the problem with jointly the government.

Second, green NGOs need to enhance their professional competency in order to achieve their goals. Scholars have pointed out that "ideals or missions are the core for NGOs' existence and development. The less satisfactory performance of green NGOs in China has much to do with their lack of ideals, missions, their lack of motivation and ambition, and their ambiguous self-identity" (Wang and Jia 2002). Chinese green NGOs must work hard to equip themselves with these qualities, which most of them are still lacking, in order to fully achieve their potential of winning the hearts and minds of the society as well as the government.

Third, Professional skills and knowledge are indispensable in solving environmental problems. Green NGOs need to improve their professional competency in solving practical environmental problems such as rubbish disposal, water pollution, and mercury pollution. As nongovernment organizations, they are expected to help tackle environmental problems with solid professional competency, down-to-earth spirit, sincere attitude, and their strong commitment in public interests. Many green NGOs still have a long way to go toward these requirements.

Fourth, instead of playing the role of passive participants in environment governance as the traditional mass organizations did under Mao and are doing even today, green NGOs should be capable of finding social issues that are well matched with their own missions and characters, and grasp opportunities to highlight their importance and strength before taking part in the network environment governance. This requires them to find appropriate strategies to expose the problem and call for action. Very often, however, some green NGOs tend to use exaggeration and rumor to draw attention, which may instead damage the reputation of green NGOs as an industry. Another common problem associated with green NGOs is their lack of discipline and organization and their being self-centered. For a better relationship and meaningful cooperation between NGOs and governments, more efforts are needed for better integration of green NGOs in order to consolidate and enhance their influence and capacity (Hong 2007).

Generally speaking, green NGOs today are more active at the end stage of an environment event. Their involvement in the early stage and in decision-making is still rare. They are usually seen as defenders of public interest only after pollution and damages have taken place. Take the PM2.5 incident for instance. Green NGOs took action on a large scale only when the pollution had deteriorated into a very bad state. Acting earlier to conduct publicity and education campaigns, as well as to make proposals for the government before the pollution turned into a nightmare, could help green NGOs to do more to alleviate environmental damages, improve the quality of environmental governance, and make NGOs themselves more acceptable to the public. It is therefore still a long way to go for green NGOs before achieving public participation in a real sense that would include multiple stages of participation, such as agenda participation, process participation, and implementation participation (Zhuang 2009). In most cases, the public has no access to the government's decision-making process about environmental issues and is unaware of the legal weapons to safeguard their rights. Green NGOs have much to offer in this regard and they need to gain more recognition and trust from the public, to expand the channels for the public to access relevant policies and regulations, and to help the public to empower themselves in solving environmental problems.

As a conclusion, the PM2.5 incident provided an opportunity for the public, especially green NGOs, to participate in environmental governance. The campaign did not entirely solve the problem of pollution and disperse the haze covering many cities in China. But the government, facing public pressure inspired by the campaign, did change its attitudes toward environmental governance and made sound progress for public involvement in environmental governance and, more importantly, recognized the importance of mutual empowerment of the government and green NGOs.

The model presented is based on a single event and needs to be applied and verified in more cases for generalization. However, it did demonstrate that the Chinese government has gradually adjusted its policies and incorporated green NGOs into its management process in order to fulfill its promise of an innovative social governance model. For that end, we expect the relationship between green NGOs and the government to have the potential to become more harmonious in addition to the progress that has already been made.

Bibliography

Chen, Zhenming. 2003. *Public Management—A Research Approach Different from the Traditional Administration*. [*gonggong guanlixue—yizhong butongyu chuantong xingzhengxue de yanjiutujing*]. Beijing: China Renmin University Press.

Hong, Dayong. 2007. *The Growth of the Chinese Private Environmental Force* [*zhongguo mingjian huanbao liliang de chengzhang*]. Beijing: Renmin University Press.

Salamon, Lester and Wojciech Sokolowski. 1999. *Global Civil Society: Dimensions of the Nonprofit Sector*. West Hartford: Kumarian Press.

Wang, Junxiu. 1994. "Social Forces in Environmental Protection: Ideals and Strategies [*huanbao shehuili: guannian ji celue*]." *Speaking and Thinking* [*yan yu si*] Vol. 4:15–24.

Wang, Ming and Xijin Jia. 2002. "An Analysis of China's NGO [*zhongguo NGO de fanzhanfenxi*]." *Management World* Vol. 8 (August):30–43.

Wang, Sining. 2008. "Practice of Cultivating the Future Green NGO Outstanding Members [*peiyang weilai luse NGO youxiuchengyuan de shijian*]." *China Out-of-school Education* Vol. 8 (August):12–16.

Zhuang, Wenjia. 2009. "The Public and Intellectual Elites in China's Environmental Governance in Transition [*zhongguo huanjinzhili zhong de shehui gongzhong yu zhshijingying*]." *Friends of Nature Communications* no.3.

Part IV

Social and Cultural Foundations of Environmental Governance

CHAPTER 9

FIRM MANAGEMENT AND
ENVIRONMENTAL ORGANIZATIONAL
VIOLENCE IN CHINA

Gary Green and Huisheng Shou

There have been strong assertions for at least the past quarter century (i.e., Hills 1987) that decisions by an organization that threaten physical harm to humans constitute "violent" behavior. Typically these harms, which we term acts of "organizational violence," derive from organizational actors' unlawful choices impacting worker safety, the physical environment, and dangerous consumer products. This chapter will put forth a general theoretical framework to explain the high rate of these phenomena in today's China, with specific focus on environmental pollution.

Our thesis is that there is an incongruously high rate of organizational violence in China's environmental pollution despite reasonable attempts to reduce it through governmental regulation, and that this state of affairs can be best understood through homogenizing theoretical frameworks from the fields of political economy and criminology. We propose that the current developmental stages of a changing China are criminogenic because they inherently facilitate organizational violence. The criminogenesis arises from the concomitant existence of two conceptually independent causes: (1) complex legal structural barriers to effective enforcement against organizational violence caused by a highly fragmented Chinese political system; and (2) a socially disorganized business environment that encourages adoption of alternative codes of conduct based on the corruptive influences of a new capitalist economy supplanting the norms of traditional institutions. We believe that the intersection of these structural and subcultural impetuses is synergistic in the exacerbation of organizational violence.

Our presentation is organized as follows: (1) operationalizing organizational violence; (2) critique of current perspectives on environmental violence in China; (3) *a priori* assumptions about profit motive and morality; (4) our explanatory model; (5) explanation of the model; and (6) conclusion.

1. OPERATIONALIZING "ORGANIZATIONAL VIOLENCE"

A strict operationalization of organizational violence is essential at the outset. Without it, we cannot know what our dependent variable comprises, and would therefore never be able to determine its causes. Stuart Hills seems to be the first to systematically make the aforementioned analogical leap to organizational "violence" when he defined *corporate violence* as "actual harm and risk of harm inflicted on consumers, workers, and the general public as a result of decisions by corporate executives or managers, from corporate negligence, the quest for profits at any cost, and willful violations of health, safety, and environmental laws" (1987:vii). Although Professor Hills deserves enormous credit for putting forth the first systematic operationalization of the idea, his conceptualization, we believe, is in need of minor revision without losing its important theoretical underpinning.

Foremost, we would agree with Hills's explicit assertion that the decisions of organizations are in all cases reducible to the decisions of individuals within them. Put another way, organizations do not behave independently of their agents (Parisi 1984; cf. Braithwaite and Fisse 1990). That stated, we would change Hills's definition only slightly by replacing the concept of "corporate" with "organizational" because the latter is all-encompassing. We acknowledge that "corporate violence" is a catchy phrase with strong political overtones that implicitly vilifies the rich and powerful as greedy and uncaring. However, "corporate" and specifically "corporate executives and managers" may easily imply that these behaviors are committed only by people in much larger organizations, or at least by those in organizations that are legally designated as corporations. Because large corporations constitute such a small proportion of all businesses, the idea of "corporate" managers and executives may implicitly exclude the countless physically harmful decisions made in the milieu of much smaller organizations, including those with few or no employees. As an example, four of every five organizations criminally convicted under US Sentencing Guidelines have fewer than 50 employees, many of which are involved in environmentally based criminal activities. Likewise, small and medium firms in China are overwhelming in number compared to large ones. For example, small and median firms counted for 99.4 percent of three million registered enterprises in 2002, and 99.88 percent of all industrial firms (Liu 2004). In many cases business behaviors that expose humans to physical harm are associated with persons in small and medium firms.

Having made this slight adjustment in defining "organizational violence," the only real argument against the concept would be that the idea of "violence" has generally been limited to direct and immediate harms intentionally inflicted upon particular individuals, such as through assault, robbery, rape, and murder, and such is not the case for the organizational behaviors with which we are dealing. The head of a company is not likely to pollute the water supply of a city in order to seek revenge on an ex-wife, for instance. Whatever persuasiveness this argument may have against the idea of organizational violence, one cannot argue that intentional choices by organizational decision makers to recklessly endanger the life and limb of other human beings shares nothing in common with assaultive behaviors that have typically been characterized as violent, especially given that single instances of organizational decisions can cause more injuries, more serious injuries, and more death than single instances of assaultive street crimes. Therefore, for the purposes at hand, it is the willful reckless physical endangerment of others shared by organizational decision makers and by individuals committing violent "street" crime that renders organizational violence to be violence *per se*.[1]

Professor Hills also, and quite rightly, includes the requirement that organizational violence should involve the breaking of law, whether criminal or regulatory. This is an important proviso because, otherwise, any physical harm that ultimately is traceable to an organizational decision could be considered organizational violence. The cases of the Ford Pinto[2] and the space shuttle Challenger disaster (Vaughan 1997) represent instances in which routine organizational processes, including "structural secrecies," can cause unintended human harm, but that does not necessarily mean the harms are the result of organizational violence. Considering any physical harm against humans based on organizational decisions to be *ipso facto* "organizational violence" makes no more sense than stating that any surgical procedure undertaken in good faith that does not result in its intended benefits constitutes unnecessary, and therefore assaultive, surgery. Further, as Perrow (1984) has told us, high-risk technologies will inevitably lead to "natural accidents," regardless of our intent to avoid them. The most important defining characteristic of organizational violence, then, is that it must involve an illegality that is intentional.

There are, of course, knotty problems that immediately arise in the operationalization of organizational violence in relation to an organizational actor's *intent* to make decisions that cause human harm. Hills's definition of violent behavior expands blameworthiness parameters to include "negligence," which is a culpability level under civil law that can be met in the event of an organization's failure to exercise "due diligence" to avoid the harm in question. Negligence may have nothing to do with intent whatsoever, especially in the cases of accidents which are, by their very nature, almost always a product of a lack of due diligence and in all cases not purposeful. Because the notion of negligence is regularly based on moral relativism and idiosyncratic fact-finding in individual cases, it is probably too elastic to be an operationalized criterion for a social science variable, and we therefore reject it.

Instead, we adopt the same criteria for intent used by the United States Sentencing Commission in their Organizational Guidelines. All organizations, regardless of their size, are expected to proactively initiate internal compliance procedures that systematically avoid crimes, torts, and other legal violations. Therefore, based on the culpability framework from the aforementioned Guidelines, we operationalize our construct "organizational violence" to include *any business behavior that participates in, condones, or demonstrates willful ignorance of a legal violation that has the potential to cause physical harm to humans*. Adopting further operationalizations from the Commission, an individual "condoned" an offense if he or she knew of the offense and did not take reasonable steps to prevent or terminate it. And, an individual was "willfully ignorant of the offense" if she or he did not investigate the possible occurrence of unlawful conduct despite knowledge of circumstances that would lead a reasonable person to investigate whether unlawful conduct had occurred (US Sentencing Guidelines, Section 8A.1.2). These parameters would include all accidents that fit (e.g., environmental spills that result from a purposeful refusal to handle toxic substances in a legally required manner, injury or death related to illegal working conditions, or accidents from dangerous consumer products where the design flaw or toxicity is known to the manufacturer or distributor or should have been known to them). Without the foundational criterion of an action violating the law and requiring the actor to participate in, condone, or be willfully ignorant of the offense, the term "organizational violence" can be used without discrimination. Further, if an agent's action is legal, it can be neither wrongful nor violent without allowing peculiar moral relativism for the definer.[3]

Our dependent variable here, then, comprises all business behaviors encompassed in our definition of organizational violence that adversely affect China's environment. Some Chinese pollution would not be included, such as freakishly accidental environmental events that are not the result of participation in, condoning of, or willful ignorance about illegal behavior. The massive oil spill that resulted from the Exxon Valdez crash near Alaska was an accident and not organizational violent behavior according to our conceptualization. However, the death of 15 workers and injury to 170 others after the explosion at the BP Texas City Refinery in March of 2005 would doubtlessly be a case of organizational violence because there were clear illegalities associated with working conditions that were ignored by BP employees and that were the cause of the disaster.

2. Critique of Current Perspectives on Environmental Violence in China

Despite what its title might imply, this chapter does not echo what the popular outcry often charges: that China lacks improvements in environmental protection or that environmental standards have actually deteriorated as China is integrated into the world economy (e.g., Economy 2004 and 2007). Quite opposite: improvements have been recognized particularly in laws and regulations. For instance, Schwartz (2008) comments that a review of China's environment laws and regulations indicates that China has developed an impressive array of tools to address environmental challenges. In many cases China surpasses World Health Organization and US standards, e.g., for automobile emissions and fuel efficiency where 2008 requirements in China exceed equivalent US requirements by 10 percent. Stalley (2010) gives detailed discussion of the improvement in environmental legal framework over the past several decades and believes that the improvement has been impressive. Stalley argues that even at the local level where Chinese local governments are notorious for ignoring the command from the central government, "there is no evidence that local governments have taken advantage of China's political decentralization to legally weaken subnational standards" (14–15).

Our focal point here, therefore, is not the temporal improvement of China's environmental protection. It is the persistent trend of organizational violence and the magnitude of the violence that has occurred despite strong legal and institutional improvements. Official documents show that the "extraordinary and severe accidents of environmental pollution"—the highest degree of environmental pollution—increased 12.3 times from 2002 to 2008, from 11 times to 135 times (Li Meng 2010). As for the individual cases, the news media frequently expose extraordinary stories where organizational persons knowingly exhibit despicable behaviors such as polluting the air with harmful materials and dumping toxic wastes into rivers on which local communities (including producers and their employees) depend. The damages are often at an extremely large scale, and firms could do so for a very long period with impunity. For instance, the chemical explosion in a plant owned by Jilin Petrochemical Corporation in 2005 caused five deaths and more than 70 injuries. The wastewater discharged from the explosion and fire extinguishment contaminated Songhua River and forced over four million people to live without running water for several days. In 2011, after news broke detailing pollution problems of the factories owned by Haerbin Pharmaceutical Group, the public was astonished to know that the leading Chinese drug manufacturer had been polluting its surrounding neighborhoods for decades, despite constant complaints from local residents. The firm had released pharmaceuticals into waterways and improperly burned medical waste without proper processing, causing serious water, land, and air contamination. The polluted air forced people in and near the factory to wear gas masks and keep their windows closed constantly.

The question therefore is—How could it be so bad for so long? The current literature has provided insufficient explanations on this incongruous phenomenon. Several explanations dominate, all of which can be seen as problematic in some way.

2.1. Limitations to Explanations Employing Capitalism

The first, economic, theory is based on a common assumption that associates the global spread of capitalism with a "race to the bottom" of government regulation (Woods 2006; Frankel 2005; Frankel and Ross 2005; DeSombre 2006; Clapp and Dauvergne 2005), especially in regard to the idea that competition for footloose capital drives governments to lower industrial standards in order to attract and retain capital. From a structural Marxist perspective, however, this argument is awkward because there are innumerable examples of concessions by capitalists to the regulations of their business sectors to ensure long-term viability of their own market share (Iversen 2005). The US Congress's passage of the Beef Inspection Act in 1906, as a reaction to the negative economic fallout of Americans reading about unsafe meat products in Upton Sinclair's *The Jungle*, is but one example (Green 1997). It is the long-term viability of a capitalist system, not the immediate negative effects on a paucity of current capitalists, that drives the discussion about who wins under new systems of regulation.

This argument also maintains that market pressures force firms to find any means to avoid regulation and public pressures because of strong competition from multinational firms. On the contrary, multinationals have been shown to have positive effects on domestic firms in a competitive market environment (e.g., Lei et al. 2005; Stalley 2010 and the chapter in this volume). More importantly, the global market pressure argument fails to explain the following question: Why do Chinese firms, being increasingly integrated into a competitive market, not care about their reputation? Both survey and case study data suggest that Chinese firms lack the motivation and even the awareness of self-regulation. In a survey of Chinese Fortune 500 companies, only 28 percent of sample firms indicated that customer demand had an influence on their environmental decisions (Lei et al. 2005). Stalley (2010) comments that "a large number of Chinese companies that escape both government and public notice feel even less market and public pressure" (47).

2.2. Limitations to the Effects of Market Pressures Resulting from Political Biases

Another reasoning behind the incongruously high incidence of environmental and other organizational violence points to imperfect market institutions in China that have deep political roots. The core of this argument is that small- and medium-sized private firms are bullied by biased governmental regulations, such as barriers to entry and various fees levied on them. Also alleged are abuses of power against private firms in various forms, including the facts that state-owned firms often monopolize markets and have better access to cheap credit from state-owned banks. Capital-poor private firms with no political protection are often financially strained to seek every means to survive, including more pollution to lower costs. Although small and medium-sized private firms in today's China can be victims of unfair competition and political powerlessness, this cannot ultimately explain their choice to commit organizational violence because financial motives to crime are not causes of decisions to commit crime (Gottfredson and Hirschi 1990:256; Hirschi and Gottfredson 2008:221). Furthermore, if the logic behind this line of argument holds, one would expect Chinese firms with more economic and political power to pollute less. In fact, many large firms in China have been equally violent. Brand-name companies such as aforementioned Jilin Petrochemical and Haerbin Pharmaceutical have enjoyed economic and political privileges to dominate the market and therefore should have been less pressured to pollute. Yet none of these firms seemed to hesitate more than any other small private firm in poisoning their consumers or devastating their surrounding communities.

Interestingly, the Western multinational firms, which may otherwise comply well in developed countries, seem also to have little regard for their reputations after they move to China. It is not uncommon to find news reports about the multinationals violating China's environmental regulations (also see Stalley in this volume). Greenpeace, a major international environmental nongovernmental organization, asserted that almost four of every five (78.6 percent) multinational firms they surveyed used double standards in China—they pollute in China but not in the developed countries. One of the most recent and notable scandals has been Apple Inc. having being named, since 2010, by Chinese environmental nongovernmental organizations (ENGOs) to be the worst in environmental protection among the major multinationals operating in China—Apple's Chinese suppliers are reported to have discharged polluted waste and toxic metals into surrounding communities severely threatening public health. Apple was also accused of being the least transparent multinational IT company operating in China in disclosing the information of its Chinese suppliers (New York Times 2011).

For both Haerbin Pharmaceutical that enjoyed political protection from local authority and Apple Inc. that did not, it is puzzling to observe large and brand-name companies that commit large-scale pollution without regard to the negative effects on their reputations. If market mechanisms do function in a way to induce the companies to care about their reputation, we should expect these two companies to have behaved differently. The fact that they pollute just like any other firm requires additional explanations if that line of thinking is to maintain credibility.

2.3. Limitations to Explanations Relying Solely on Low Enforcement Levels

Other common perspectives in political economy that use insufficient governmental regulatory enforcement as the major factor promoting business decisions that risk human harm (in China and elsewhere) surely hit the nail more squarely on its head by focusing on an obviously essential factor affecting individuals' proclivities to make physically harmful business decisions. Indeed, despite China's relatively high degree of regulatory laws protecting the environment, implementation does present a serious challenge. Weak enforcement is probably the most frequently cited problem in both academic research and popular discussion.

However, even if enforcement was much stronger, that does not mean that the rate of organizational environmental violence would significantly decrease. First, many may not be deterred because any threatened punishment would be beyond their abilities to pay the financial penalty, thereby eliminating any additional perceived sanction threat associated with both higher fine levels and higher violator capture rates—known as "the deterrence trap" (Coffee 1981; Green and Bodapati 2000). Second, those contemplating violation of laws may simply believe that the gains of the offense would outweigh any nebulous certainty associated with the punishment, regardless of the threatened severity of punishment. Most importantly, good enforcement does not necessarily guarantee good compliance if firms lack the incentives to self-police. As political economists have well explained, absent these incentives, enforcement by the Chinese central government can be very costly and ineffective, and eventually subject to failure (e.g., Zhang 2010). Our point in questioning the efficacy of a pure economic model of regulatory deterrence, then, is that any success associated with threatened sanctions simply masks or distorts persons' true propensities to commit organizational violence (Hirschi and Gottfredson 2008:220). We aim to theorize about the ways in which we can reduce altogether propensities to commit organizational violence as opposed to temporarily suppressing them with legal threats. We therefore argue that we must look deeper into the social normative systems that affect the morality of people in order to understand what makes firms less willing to comply and self-regulate in environmental governance, regardless of the level of enforcement.

3. *A PRIORI* ASSUMPTIONS ABOUT PROFIT MOTIVE AND MORALITY

3.1. Profit Motivation Does Not Explain Organizational Violence

The discussion above suggests an individualist approach for better understanding the recurrence of organizational violence in China's current environmental situation. In the questions raised earlier: "Why do Chinese firms lack self-enforcement and care less about their consumers' well-being?" our individualist approach means that the focus is on "managers" rather than "firms" because it is fundamentally up to the individuals who are in charge of the firms to make decisions concerning pollution. Thus, we assume that all actions of an organization are committed by its agents. This is an important departure from the existing literature because previous studies tend to transcend individual decision makers by focusing on firms as the unit of analysis and reference. Once firms are taken as the unit of analysis, individuals' preferences and choices become more irrelevant to the theorizing. Moreover, such an approach allows individuals, as agents, to easily far-distance themselves from the decisions they make by blaming the system or environment in which their firms are located or based, while at the same time the separation between firms and managers leads to a dilemma facing corporate regulation and punishment, because firms become an entity that have, as Coffee (1981) called it, "no soul to damn, no body to kick."[4]

The implications of such an approach are significant. It assumes that certain macro-level entities such as firms can corrupt an otherwise innocent individual. Good people are forced to commit crime, as suggested in the previous discussion that blames political biases against small and medium private firms for individuals' choices to commit environmental violence. Markets give poor individuals opportunities—or lure them—to improve their financial situation, while a harsh and unfair political system forces them to achieve this goal using less decent means. Such an assumption well echoes the influential "strain" theory (Merton 1938), which argues that social structures that emphasize financial success may ultimately pressure individuals to "innovate" criminal behaviors to achieve those goals when opportunity for legitimate attainment of those goals is structurally blocked. This pressure, or strain, is caused by the discrepancy between culturally defined goals and the institutionalized means available to achieve them. When material success, for example, becomes the dominant cultural goal in a booming economy like China, yet the legitimate means to achieve this goal, such as education and access to fair markets, are not uniformly distributed (e.g., those from wealthier backgrounds have considerably more access to these means than do those who are economically and politically disadvantaged), strain is generated and produces certain coping strategies for the disadvantaged. Illegal behavior is among the strategies used by the disadvantaged to deal with the pressures that are brought to bear on them.

However, each individual's perception of financial success is a matter of *relative* disadvantage and deprivation. A person who is in poverty is in absolute deprivation and seeks merely to achieve the lower end of the next highest economic class and thereby be able to enjoy the basic necessities of life. Those who are in the middle class and have achieved the basic necessities, and then some, may also feel relatively deprived because they want a nicer place to live and fancier cars and clothes. Even a person who is extremely wealthy relative to virtually everyone else may also still believe he or she has not yet reached the financially prescribed societal goal of acquisition sufficiently. Thus, "strain" is not based on absolute deprivation. It is relative to the perceptions of each individual. This explains why businesspersons who are clearly successful by most measures still choose to commit organizational violence. There is theoretically no endpoint to one's desires for additional financial success.

Despite the great attention over many decades that has been given to perceived strain as a cause of criminal behavior, we reject the idea that material success as a socially defined goal should be taken as an explanatory variable for individuals' behaviors within firms. As we have noted, motives are not causes, and therefore the seeking of profit cannot be a cause. Criminological theory requires "that crime be understood without reference to motives and benefits" (Hirschi and Gottfredson 2008:221). Profit, as motive, is a constant that applies to every rational individual and therefore cannot explain the variation in criminal behavior among individuals. As Sutherland (1947:7) simplistically and cogently argues, while criminal behavior is a means to meet financial success goals, it is not explained by those goals because noncriminal behavior is an expression of the same goals. Put differently by Sutherland (1973:39), "People steal [for various reasons] . . . . and they engage in lawful employment [for the same] reasons." Merton (1938) himself identifies multiple "modes of adaptation" to strain, and not all of them involve criminal behavior. Pressed by strain notwithstanding, one can simply remain at their current financial level or continue to work toward upward financial mobility by legal means. Put differently, individuals always have multiple options in life other than poisoning neighbors or maiming workers and consumers.

3.2. Morality and Organizational Violence

According to Hirschi (1969; also Gottfredson and Hirschi 1990), the most obvious influence on an individual's decision to commit organizational violence is their level of attachment to the feelings of other human beings. We therefore need to conceptually eliminate from our explanation, *a priori*, those with strong moral attachment to the feelings of others and a clear social bonding that ties them to their community. They are highly unlikely to engage in organizational violence (although not infallible), they need not be deterred away from business decisions involving violence, and they are essentially immune to the motivation to violence inherent in socially prescribed financial success goals. This leaves us with those who lack sufficient moral attachment to the feelings of others and who, unless otherwise deterred, may choose to utilize various illegal means to achieve their financial goals, including organizational violence. Therefore, it is only those without adequate attachment that are relevant to theorizing about the factors that promote choices to commit organizational violence. From the perspective of morality, a firm manager who decides to dump toxic wastes into a river that is critical for the surrounding community must be capable of seeing the pain his or her action is inflicting on the humans that will be adversely affected to be less important than the financial or other gain accrued from such acts of violence.

We further assume that individuals are highly unlikely to switch from a moral group to an immoral one, or vice versa. By the time people are in a position to make the choice, either they are adequately socialized against committing organizational violence or they are not. We believe that such socialization is generationally linked in that the generation dominating today's Chinese business environment has a higher propensity to lack moral attachment to the feelings of others than the previous generations that were more socialized. It is certainly subject to debate whether individuals under Mao were more or less attached to the feelings of others than today's generations.[5] The point here is that the commercialization China has undergone in the past few decades has significantly damaged social bonds—if any were left from Mao's era—that otherwise would be important in curbing the violent behavior under discussion.

The amount of organizational violence that will be perpetrated by those who are morally unopposed to it will vary based on the influences of threats of punishment and social pressures. Put another way, those who are morally capable of organizational violence can be deterred even though they still maintain a lack of attachment to the feelings of others. Deterrence merely hides or distorts downward their true propensities to commit organizational violence. The best way to reduce immoral individuals' propensity to commit organizational violence is to moralize them into believing the wrongfulness of such behavior, but that is generally unlikely given the stability of our moral systems from early adolescence through adulthood (Gottfredson and Hirschi 1990). Thus, our best prospect for reducing this type of violence is with the upcoming generation of young persons who can be morally socialized about the unequivocal wrongfulness of such behaviors through the traditional socialization avenues of schools, parents, peers, and various forms of media.[6]

4. AN EXPLANATORY MODEL OF ORGANIZATIONAL VIOLENCE IN CHINA

Crime, including organizational violence, is most likely to occur in social systems that are disorganized. Social disorganization is defined as a lack of widespread consensus about conduct norms (Sutherland 1949) and manifests under two main conditions. The first is *anomie*, or a lack of norms altogether, and is most likely to occur when a social system is in transition from one set of norms to another, such as in today's China. The postreform legal norms have not yet taken hold as China moves through a rapidly expanding capitalistic economic system. The second type of social disorganization is "differential social organization," which refers to a conflict in conduct norms that exist in a social system—where pockets of subculture are organized around sets of anti-legal conduct norms or otherwise different from those of the larger society. Our model assumes that both types of social disorganization—anomie and differential social organization—are in operation in today's China.

In Figure 9.1 we hypothesize two explanatory variables that affect the level of organizational violence, and conspicuously eliminate profit motive as a cause. The first explanatory variable is the level of regulatory and other legal enforcement, and is inversely related to the rate of organizational violence. The second explanatory variable is the level of social disorganization, which has a positive relationship with the rate of organizational violence. We also assert a bidirectional negative relationship between the level of enforcement and the level of social disorganization. This bidirectional negative relationship creates a mutual interaction between the explanatory variables that can either exacerbate the problem of organizational violence or drastically improve it. Thus, each explanatory variable exerts an independent effect on the rate

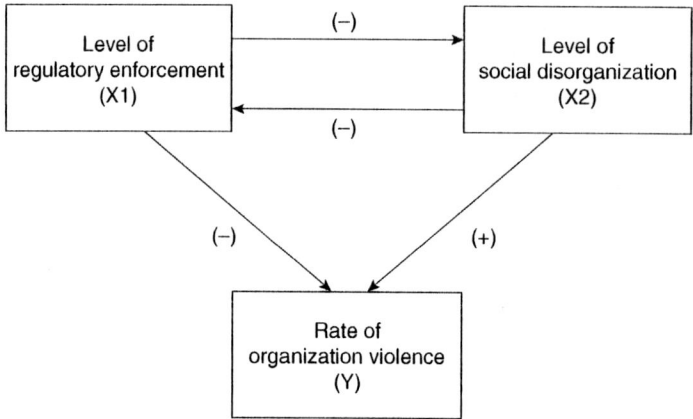

Figure 9.1 Explanatory model of organizational violence in China's environmental governance
Note:
(+) Positive Relationship.
(−) Negative Relationship.

of organizational violence, and they interact with each other in negative relationships. We explain the model in the next section.

5. Explanation of the Model

5.1. Levels of Enforcement Negatively Affect Rates of Organizational Violence

As noted earlier, many have argued that a lack of deterrence based on inadequate formal governmental enforcement of environmental rules plays an important role in individuals' decisions to commit organizational violence. We, of course, agree that governmental punishment, or more accurately the lack of it, has a fundamental influence on individuals' perceptions and strategies about risks and benefits and, as a result, their behavior. But exactly how do current political structural arrangements in China minimize the deterrence of environmental violence?

The most noticeable characteristic of the Chinese political structure concerning the implementation of public policies and laws is its fragmented nature. The influential model of "fragmented authoritarianism" (Lieberthal 1995; Saich 2004) explains that, on the surface the Chinese state is powerful in implanting laws and regulations passed down from the center to various layers of subnational territorial units—province, prefecture, country, township, and village. This is made possible through a complicated hierarchical system in which the governmental functional units, such as the Ministry of Environmental Protection (MEP), are replicated down through each of the lower territorial units, and the local environmental protection bureau (EPB) at each level is responsible to the higher level of environmental administration. In reality, however, local EPBs are greatly constrained in implementation by various actors.

Local EPBs are embedded in local governments and are often subordinate to and dependent on the local governments. The lion's share of local EPB's budget comes from the local government rather than from higher levels of the environmental protection administration. In addition, at any level of government, multiple units have competing bureaucratic interests. For example, at the center, Ministry of Land and Natural Resources, Ministry of Construction, Ministry of Water Resources, and National Development and Reform Commission, among others, may be involved in the environmental issues and the MEP has never been powerful enough to compete with these agencies, despite the fact that in recent years it has gained substantial power. Local EPBs face the same, if not worse, situations in competing with local agencies.

Local EPBs are also constrained by the intervention of the local branch of the Communist Party at each level. Furthermore, because a similar situation exists for local courts relative to local governments and Party branches, local EPBs often find it difficult to use courts to defend their action against environmental violators, either coming from the higher level or occurring at the local level.

Such a "complex arrangement of diffuse agencies" (Jahiel 1998) in the Chinese political system creates "a situation in which there is no single set of environmental regulators, but multiple bureaucratic agents with overlapping authority and divergent interests" (Stalley 2010:32). That is the primary factor contributing to the low capacity of the environmental regulators, which lack funding, personnel, equipment, training, and above all, the bargaining power against other government agencies (Li 2008; Stalley 2010; Carter and Mol 2007; Schwartz 2008). The result is an "enforcement trap" as Stalley (2010) describes—despite China's centralized creation of a quite impressive legal and regulatory system protecting the environment, and in some cases China even sets their standards beyond the international level (Schwartz 2008), enforcement of these standards is missing.

Thus, the fragmented political system renders the regulatory regime in China selective, incomplete, and, as a result, ineffective. Absent a perception of punishability among those contemplating organizational violence, the system fails to deter. Individuals may, as numerous cases have illustrated, instead find the risks of being caught and then fined to be outweighed by the benefits from committing violence. Ironically, the lack of effectiveness in the regulatory regime promotes governments to use political means to supplement the system. Mass publicity campaigns have been routinely used to deter the wrongdoing through single harsh punishments. Not only does singling out for deterrence create punishments that are intemperate (precisely because the purpose of the punishment is to draw attention to it to set an example), but such harsh punishments ironically often result in hollow rhetoric without lasting substantive compliance. Oftentimes things return to normal as quickly as the campaign storms come, further convincing individuals to take risks.

5.2. Levels of Enforcement Negatively Affect Social Disorganization

But the effects of inadequate enforcement on the rate of organizational violence go far beyond a lack of deterrence because, as we noted earlier, ratcheting up enforcement may not result in a commensurate level of legal compliance. As Figure 9.1 suggests, political structure affects individuals' behavior not only through deterrence or lack of it, but also because it affects behavior indirectly by shaping the social environment that would provide informal rules to regulate social members through norms and social networks. That is, the social environment is to a large extent a function of the recursive messaging associated with the formal enforcement process. The lack of enforcement, therefore, fails to combat the growth of subcultures that subscribe to alternative moral codes of conduct. Put another way, the absence of official governmental recursive messaging against environmental (and other forms of) organizational violence allows unhealthy levels of immorality to develop that embody the opposite of legal compliance. This differential social organization is a direct product of the lack of enforcement caused by the fragmented political structure, and is conceptually quite distinct from individual-level deterrence.

This means that the lack of enforcement caused by the structural problem in China's political system retards Chinese society from producing normative structures against violent behavior. Put in the longstanding terms of sociology and political economy, informal institutions that are manifested in customs, traditions, codes of conduct (norms and rules), and other areas of informal social control that are parallel to formal institutions (such as laws and regulations) serve to maintain social order, and to sanction and normalize social behaviors (North 1990; Ostrom 1990). Their current absence in China constitutes the first part of the bidirectional negative relationship between the explanatory variables that we assert—the lack of formal punishment based on the fragmented political structure hinders the development of informal social controls against organizational violence, thereby encouraging social disorganization.

5.3. Social Disorganization Negatively Affects Levels of Enforcement

The second half of the bidirectional relationship is the inverse causal ordering of the first—societal/informal institutions underpin the function of formal institutions in that social norms determine whether formal institutions (including enforcement levels) can be implemented as expected. Formal institutions created from the top often encounter the resistance at the local level of a society, which holds a set of different social norms (i.e., that is socially disorganized) (Migdal 1988; Migdal, Kohli and Shue 1994). In other situations, suitable social norms (a lack of social disorganization) can facilitate the implementation of formal rules, as discussed widely in the literature associated with social capital (Becker 1996; Coleman 1988; Putnam 2000; Evans 1997; Ostrom 1997).

China fails to hold common codes of conduct against organizational violence for three reasons. First, the political structure that fails to punish the violence will create a situation in which social norms are absent in constraining, or too fragmented to respond to, violent behavior. A transitional society is often subject to such a situation where a shift in traditions and values creates social turmoil because the old norms have not yet been replaced by new ones (i.e., there is "anomie" or normlessness). Further, as O'Donnell (1993, 1994) points out, what aggravates the situation are often the political forces such as the abuse of power by officials that may lead to "the angry atomization of society" that cultivates distrust and cynicism among citizens toward the public authority and public good such as regulation, compliance, and social responsibility. This not only jeopardizes the regulatory regime top-down, but also a skeptical and cynical public is a weak public in controlling itself from the bottom-up. In other words, lacking the normative structure against violent behavior undermines and weakens the formal legal social control functions of the Chinese government.

This also at least partially explains the weak influence of environmental groups on the public. The current environmental protection organizations are often elitist, popular among the newly emerged middle class and college students but have little appeal to the public. That is not entirely the result of political restriction but is closely associated with the lack of efficacy among the public regarding environmental protection as a public good. It is therefore common to find the public complaining without actions to follow. As Ho (2007) argues, it is too early to claim that a Chinese environmental movement is in place. The public often easily forgets and sometimes even forgives those firms that have caused horrifying pollution. As evidence, firms such as the aforementioned Jiling Petrochemical and Haerbin Pharmaceutical, after a short period of media exposure and governmental investigation, continued their polluting practices without consumer outcry or boycott.

Second, in a society without a normative structure, individuals may resort to their private authorities that may permit or even encourage violent pursuit for profits. Sutherland's (1949) monograph *White Collar Crime* used his theory of differential association to account for organizational wrongdoing. According to Sutherland, criminal behavior is learned in interaction with other deviant persons. Through this association, they not only learn the specific justifications and other attitudes favorable to violation of law, but they also learn criminal techniques. Sutherland asserts that a person becomes deviant because of an excess of definitions favorable to violation of law over definitions unfavorable to violation of law. In other words, criminal behavior emerges when one is exposed to more messages favoring criminal conduct (including failure to demonize it through meaningful formal legal enforcement mechanisms) than pro-legal messages. As evidence, organizational violence in China is often clustered within particular regions or industries that can easily share information or emulate. Meanwhile, in a situation like today's China, individuals are more likely to learn criminal behavior not only because of the easy exposure to and contact with criminal patterns but also because of their isolation from anticriminal patterns.

Alternatively, persons can simply invent ways to commit organizational violence as well as methods to avoid detection for committing it, but the more sophisticated the *modus operandi*, the more likely is the necessity of learning it from other offenders. Anti-legal conduct norms and emulation of techniques of organizational violence will continue to spread through contagion to the extent those subcultures are not counteracted by pro-legal messaging.

Third, the lack of punishment in the political structure and the lack of social control function in the society, working together, weaken the society's ability to defeat profit as a culturally accepted goal that dominates the society. Profits often supersede morality in individuals' calculation and decision making, which in effect erodes social bonds. Once the moral links to other people are undermined and weakened, it is not surprising to observe individuals perpetrating violence for quick financial success without much sanction from a society that is fragmented.

In sum, the current social environment in China lacks normative validation—the socialization factor associated with recursively messaging the inappropriateness of organizational violence through *actual* punishment inflictions. This situation creates a public that is incapable of controlling its social members and, consequently, fails to sanction a socially disorganized business culture that encourages adoption of alternative codes of conduct based on the corruptive influences of a new capitalist economy supplanting the norms of traditional institutions. This downward spiral of immorality will continue in the absence of much stronger and highly continuous social and legal messages that promote attachment to others individually and communally. This is especially true for the up-and-coming generations.

6. Conclusion

Today's China is a combination of a booming economy, a mobile society, a young population, and an ineffective political system—a "perfect storm" for the facilitation of organizational violence. In subordinating morality for profit, Chinese businesspersons commonly take Merton's "innovation" as their mode of adaptation: using illegitimate means to pursue the culturally pressured financial success goals. This is a familiar situation that occurred in the United States during the gilded era from the late nineteenth century up to the early twentieth century (Messner and Rosenfeld 1994). As history demonstrated, combating it was not an easy task. It took decades and concerted efforts from both the top and bottom, with the assistance of a major economic crisis and a World War, to curtail it in the United States. China presents additional difficulties of fragmented political structure and a lack of formal channels for public involvement, suggesting a much more difficult future to build an effective regulatory regime and a society morally opposed to organizational violence.[7]

Yet it must be clear that this trend is not insurmountable. It is fundamentally determined by individuals, and by individuals' willingness to accept it or oppose it. Thus we reject a deterministic view. In fact, some recent signs suggest that a movement is underway to reverse this trend.

The government has been strengthening the capacity of environmental administrations in reforming its regulatory regime, as Brettell's chapter in this volume has suggested. It has clearly prioritized environmental protection and taken it to key levels of official evaluation and implementation. The government has also been encouraging public hearings in the decision-making process concerning the environment. Market forces have also contributed positively to this swing. China's further integration into the world economy implies more liberalization and more pressures of competition on firms from both domestic and international markets. Upgrading skills has been an inevitable task for domestic firms to survive, suggesting less pollution for producing value-added products. Socially, ENGOs are quickly expanding despite the political barriers, and the newly emerged middle class has gained influence through new technologies associated with internet. New resources and strategies are emerging.

These all suggest that a disordered culture lacking normative structure against organizational violence is under challenge and a new consensus for humane business conduct norms is being slowly erected in China, though it is far from clear whether a tipping point has been reached or will be reached soon.

Notes

1. Salmi (2004:56–57) has differentiated between "direct" and "indirect" violence, and organizational violence in our context would fall as a subcategory of the latter because it is a "result of a deliberate human intervention in the natural or social environment whose harmful effects are indirect or often delayed."

2. A classic example would be the long-time wrongful designation as "violent" Ford Motor Company's actions in regard to its marketing of the Pinto in the late 1960s and early 1970s (e.g., Green 1997:129–130), even though it knew that the vehicle's gas tank mounted behind the rear axle could immeasurably increase the probability of fiery explosion as the result of a rear-end collision. Indeed, Ford's now-famous Grush–Saunby memorandum, which juxtaposes the costs of retooling (to relocate) the gas tank relative to the probable civil settlement payments in burn injury and death lawsuits, has been seen as the epitome of organizational amoral profit rationality. However, as Lee and Ermann (1999) convincingly demonstrate, the Grush–Saunby memorandum occurred in 1973 (long after the Pinto had been marketed), that it was generated for the National Highway Safety Traffic Administration (NHSTA) in the negotiation of a possible safety standard and not internal distribution among Ford decision-makers, that it was based on NHSTA accepted procedures for cost–benefit analyses in such matters, that crash-testing was not an established practice in the industry at the time Pinto was designed and manufactured for several years, and that the Pinto's questionable design was quite legal because it did not violate any regulatory law. Thus, it is difficult to attach a label of "violent behavior" to those involved, especially when there is no evidence of

purposeful action to break the law. In fact, Ford was concerned about the issue and took proactive measures to investigate it (Lee and Ermann 1999).
3. Our conceptualization of organizational violence can potentially be criticized for its speciesist bias (Beirne 1999)—only human animals can be victims of it by our definition. Surely, non-human animals are also victims of organizational violence, such as fish that die in polluted rivers, dogs and cats that die from poisoned pet food consumer products, and work animals that are harmed or killed based on illegal cruel working conditions. However, we think it wise at this point to limit victimization from organizational violence to human beings, but are nevertheless cognizant of the meaningfulness of a non-speciesist conceptualization as well.
4. For a general theoretical discussion about the relative importance of the individual versus firm level in the explanation of offending within organizations, see Reed and Yeager (1996), Herbert, Green, and Larragiote (1998), and Yeager and Reed (1998).
5. For discussion on the communitarian tradition of social control under Mao and the change after, see, e.g., Rojek (2001), Deng and Cordilia (1999), Lu and Miethe (2001); Liu (2006); Zhang Jiang and Lambert (2009).
6. See Liu, in this volume, on this matter.
7. On China's social control and crime during transition, see Liu, Zhang and Messner (2001); Liu (2009); on corporate punishment in today's China, see Zhang and Zhao (2007).

References

Becker, Gary S. 1996. *Accounting for Tastes.* Cambridge, MA: Harvard University Press.
Beirne, Piers. 1999. "For a Nonspeciesist Criminology: Animal Abuse as an Object of Study." *Criminology* 37: 117–147.
Braithwaite, John and Brent Fisse. 1990. "On the Plausibility of Corporate Crime Theory." *Advances in Criminological Theory* 2: 15–38.
Clapp, Jennifer and Peter Dauvergne. 2005. *Paths to a Green World: The Political Economy of Global Environment.* Cambridge, MA: MIT Press.
Coffee, John Jr. 1981. " 'No Soul to Damn; No Body to Kick': An Unscandalised Inquiry into the Problem of Corporate Punishment." *Michigan Law Review* 79: 386–459.
Coleman, James S. 1988. "Social Capital in the Creation of Human Capital." *American Journal of Sociology* 94: 95–120.
Deng, Xiaogang and Ann Cordilia. 1999. "To Get Rich is Glorious: Rising Expectations, Declining Control, and Escalating Crime in Contemporary China." *International Journal of Offender Therapy and Comparative Criminology* 423: 211–229.
DeSombre, Elizabeth. 2006. *Flagging Standards: Globalization and Environmental Safety, and Labor Regulations at Sea.* Cambridge, MA: MIT Press.
Economy, Elizabeth. 2004. *The River Runs Black: The Environmental Challenge to China's Future.* Ithaca, NY: Cornell University Press.
Economy, Elizabeth. 2007. "Environmental Governance: The Emerging Economic Dimension." In Neil Carter and Arthur Mol (eds.) *Environmental Governance* in *China.* Rutledge: London and New York.

Evans, Peter. (ed.) 1997. *Stare-Society Synergy: Government and Social Capital in Development*. Berkeley: University of California Press.
Frankel, Jeffrey. 2005. "The Environment and Economic Globalization." In Michael Weinstein (ed.) *Globalization: What's New?* New York: Columbia University Press.
Frankel, Jeffrey and Andrew Ross. 2005. "Is Trade Good or Bad for the Environment? Sorting Out the Causality." *Review of Economics and Statistics* 87(1): 85–91.
Gottfredson, Michael R. and Travis Hirschi. 1990. *A General Theory of Crime*. Stanford: Stanford University Press.
Green, Gary S. 1997. *Occupational Crime* (2nd edition). Chicago, IL: Nelson-Hall Publishers.
Green, Gary S. and Madhu Bodapati. 2000. "The Deterrence Trap in the Federal Fining of Organizations." *Criminal Justice Policy Review* 10(4): 547–559.
Herbert, Carey, Gary S. Green and Victor Larragiote. 1998. "Clarifying the Reach of *A General Theory of Crime* for Organizational Offending: A Comment on Reed and Yeager." *Criminology* 26(4): 867–883.
Hills, Stuart (ed.). 1987. *Corporate Violence: Injury and Profit for Death*. Totowa, NJ: Rowman and Littlefield.
Hirschi, Travis. 1969. *Causes of Delinquency*. Berkeley: University of California Press.
Hirschi, Travis and Michael R. Gottfredson. 2008. "Critiquing the Critics." In Erich Goode (ed.) *Out of Control*. Stanford: Stanford University Press.
Ho, Peter. 2007. "Embedded Activism and Political Change in a Semi-authoritarian Context." *China Information* XXI(2): 187–209.
Iversen, Torben. 2005. *Capitalism, Democracy, and Welfare*. Cambridge: Cambridge University Press.
Jahiel, Abigail. 1998. "The Organization of Environmental Protection in China." *China Quarterly* 156: 757–786.
Jiang, Shanhe and Eric G. Lambert. 2009. "Views of Formal and Informal Crime Control and Their Correlates in China." *International Criminal Justice Review* 19(1) (March): 5–24.
Lee, Matthew, and David Ermann. 1999. "Pinto 'Madness' as a Flawed Landmark Narrative: An Organization and Network Analysis." *Social Problems* 46(1): 30–47.
Lei, Peng, Baijin Long and Pamlin Dennis. 2005. "Chinese Companies in the 21st Century." WWF's Trade and Investment Programme. http://www.wwf.se/source.php/1158886/chinese_companies_in_the_21st_century_bl.pdf
Li, Meng. 2010. "Distorting Behavior of Local Governments in Environmental Governance [*defang zhengfu zai huanjing zhili zhongde niuqu xingwei*]." *Environmental Protection* [huanjing baohu] 13: 23–26.
Li, Wanxin. 2008. "China's Environmental Regulation and Governance: Ideas, Promises, Capacity, and Endowments [*zhongguo de huanjing jianguan yu zhili*]." *Journal of Public Administration* [*gonggong xingzheng pinglun*] 5: 102–151.
Lieberthal, Kenneth. 1995. *Governing China: From Revolution to Reform*. New York: W. W. Norton & Company.
Liu, Jianhong. 2006. "Modernization and Crime Patterns in China." *Journal of Criminal Justice* 34(2) (March–April): 119–130.
Liu, Jianhong, Lening Zhang and Steven Messner (eds.) 2001. *Crime and Social Control in a Changing China*. London: Greenwood Press.
Liu, Qiang. 2004. "Situation of and Responses to of Regulating Work Safety in Small and Median Enterprises." *Labor Protection* [*laodong baohu*] 1: 5–13.

Lu, Hong and Terance Miethe. 2001. "Community Integration and the Effectiveness of Social Control." In Liu, Jianhong, Lening Zhang and Steven Messner (eds.) *Crime and Social Control in a Changing China*. London: Greenwood Press, pp. 105–122.
Merton, Robert K. 1938. "Social Structure and Anomie." *American Sociological Review* 3: 672–682.
Messner, Steven and Richard Rosenfeld. 1994. *Crime and the American Dream*. Belmont: Wadsworth.
Migdal, Joel. 1988. *Strong Societies and Weak States: State-Society Relations and State Capabilities in the Third World*. Princeton, NJ: Princeton University Press.
Migdal, Joel, Atul Kohli and Vivienne Shue. 1994. *State Power and Social Forces: Domination and Transformation in the Third World*. Cambridge: Cambridge University Press.
Neil, Carter and Arthur Mol (eds.). 2007. *Environmental Governance in China*. Routledge: London and New York.
New York Times. 2011. "Apple Cited as Adding to Pollution in China." September 2.
North, Douglass. 1990. *Institutions, Institutional Change and Economic Performance*. Cambridge: Cambridge University.
O'Donnell, Guillermo. 1993. "On the State, Democratization and Some Conceptual Problems: A Latin American View with Glances at Some Postcommunist Countries." *World Development* 21(8): 1355–1369.
O'Donnell, Guillermo. 1994. "Delegative Democracy." *Journal of Democracy* 5(1): 56–69.
Ostrom, Elinor. 1990. *Govern the Commons: The Evolution of Institutions for Collective Action*. Cambridge: Cambridge University Press.
Ostrom, Elinor. 1997. "Crossing the Great Divide: Corproduction, Synergy, and Development." In Peter B.Evans (ed.) *State-Society Synergy: Government and Social Capital in Development*. Berkeley, CA: International and Area Studies, pp. 85–118.
Parisi, Nicolette. 1984. "Theories of Corporate Criminal Liability." In Ellen Hochstedler (ed.) *Corporations as Criminals*. Beverly Hills, CA: Sage Publications.
Perrow, Charles. 1984. *Normal Accidents: Living with High-Risk Technologies*. Princeton, NJ: Princeton University Press.
Putnam, Robert. 2000. *Bowling Alone: The Collapse and Revival of American Community*. New York, NY: Simon and Schuster.
Reed, Gary E. and Peter C. Yeager. 1996. "Organizational Offending and Neoclassical Criminology: Challenging the Reach of *A General Theory of Crime*." *Criminology* 34: 357–382.
Rojek, Dean. 2001. "Chinese Social Control: From Shaming and Reintegration to 'Getting Rich Is Glorious.'" In Liu, Jianhong, Lening Zhang and Steven Messner (eds.) *Crime and Social Control in a Changing China*. London: Greenwood Press, pp. 89–104.
Saich, Anthony. 2004. *Governance and Politics of China*, 2nd Edition. New York: Palgrave Macmillan.
Salmi, Jamil. 2004. "Violence in Democratic Societies: Toward an Analytic Framework." In Hillyard, Paddy, Pantazis Christina, Tombs Steve and Gordon Dave (eds.) *Beyond Criminology? Taking Harm Seriously*. London: Pluto Press, pp. 55–66.
Schwartz, Jonathan. 2008. "China's Environmental Policy and Activities to Address the Environmental Impacts of its Energy Use." Testimony before the U.S.-China Economic and Security Review Commission. The text is

available on http://www.uscc.gov/Hearings/hearing-chinas-energy-policies-and-their-environmental-impacts

Stalley, Phillip. 2010. *Foreign Firms, Investment, and Environmental Regulation in the People's Republic of China*. Stanford, CA: Stanford University Press.

Sutherland, Edwin. 1947. *Principles of Criminology*, 4th ed. Philadelphia: Lippincott.

Sutherland, Edwin.1949. *White Collar Crime*. New York: Holt, Rinehart & Winston.

Sutherland, Edwin. 1973. "Critique of the Theory." In Karl Schuessler (ed.) *Edwin H. Sutherland: On Analyzing Crime*. Chicago: University of Chicago Press, pp. 13–29.

Vaughan, Diane. 1997. *The Challenger Launch Decision: Risky Technology, Culture, and Deviance at NASA*. Chicago: University of Chicago Press.

Woods, Neal. 2006. "Interstate Competition and Environmental Regulation: A Test of the Race-to-the-bottom Thesis." *Social Science Quarterly* 87(1): 174–189.

Yeager, Peter Cleary and Gary E. Reed. 1998. "Of Corporate Persons and Straw Men: A Reply to Herbert, Green, and Larragiote." *Criminology* 36: 885–898.

Zhang, Ling and Lin Zhao. 2007. "The Punishment of Corporate Crime in China." In Henry N. Pontell and Gilbert Geis (eds.) *International Handbook of White-Collar and Corporate Crime*. Heidelberg: Springer, pp. 663–679.

Zhang, Weiying. 2010. *Logic of Markets [shichang de luoji]*. Shanghai: Shanghai People's Press.

CHAPTER 10

SOCIALIZATION AND
INTERGENERATIONAL CHANGE OF
ENVIRONMENTAL CONSCIOUSNESS
IN CHINA

Xiaoqing Liu

1. INTRODUCTION

Public participation is an important part of the mechanism of dealing with environmental protection. Different from the bottom-up model in environmental mobilization and governance in the West, China follows a top-down model in which the government makes decisions, implements policies, and educates the public. Ever since the beginning of economic take-off, the Chinese government has made environmental protection a national policy. In 1992, the government proposed a strategy for sustainable development, and shortly after issued *China 21st Century Agenda*, for the first time advocating public participation. With such political determination (Pan 2004), the public environmental participation system was established gradually. It is therefore safe to say that China's public environmental participation system was characterized by governmental dominance.

It is therefore natural to question about how effective this state-led system is in cultivating the public consciousness of environmental participation after 20 years of development. Three questions need to be answered. First, has the Chinese pubic cultivated consciousness of environmental participation after two decades of government guidance? Second, how unique is the process of China's socialization in environment protection? Third, how does this process influence the public consciousness of environmental participation?

To answer these questions, this chapter will analyze public preference for environmental governance using data from two nationwide surveys. Public preference for environmental governance offers a useful perspective to not only understand the public consciousness of environmental participation but also evaluate the performance of the government-led environmental protection system.

2. Research Design and Variable Description

2.1. Hypotheses

The process of political socialization is a process of interaction between individuals and political system, through which the political culture is cultivated among the public. Since China's environmental protection system has been led by the government in the past 20 years, the process is the one through which the government propagates its ideas and ideals while individuals accept. Although families, schools, peer groups, work units, and mass media are all important agents of the government in this process (Moschis and Churchill 1978), under the current government-led public participation system, propaganda and mass education are the main tools of social mobilization. Geping Qu, the then director of the State Environmental Protection Bureau, pointed out in 1988 that "the reason why we attach importance to propaganda education is determined by our national realities. The history of environmental protection in China is very short. So, both the cadres of all levels and the public have limited consciousness on the importance of this matter. Besides, our culture and technology are both backward, making it more difficult for the public to improve the environmental consciousness and morality autonomously" (National Environmental Protection Bureau 1988). With this recognition, the environmental protection has entered into all levels of educational programs including colleges and primary schools. This means the school is the main agent in political socialization of environmental protection.

From life course theory of political socialization, the teenage period is critical for a citizen to acquire political civilization and form political attitudes and tendencies (Weissberg 1974). That means citizens who receive government's environmental education, especially during their teenage years, may have a stronger awareness about environmental participation than others.

Based on these analyses, this chapter puts forward the following hypothesis: intergenerational difference is the most critical factor affecting public preferences for environmental governance. To test the hypothesis and calculate the net influence of this factor, some control variables are included. According to environmental consciousness theory, the public preferences for environmental governance are affected by social and economic status (SES), environmental perception, and experience of pollution. According to the governance theory, public participation is a remedy to the deficiency of government and market mechanisms. The public evaluation of local government performance therefore should be taken into consideration as a factor affecting the public participation into local environmental affairs. To summarize, control variables in this study will include SES, environmental perception, pollution experience, and government performance evaluation.

2.2. Descriptive Statistics

2.2.1. Dependent Variable: The Public Preferences for Environmental Governance

When asked in the 2008 survey about their opinions on the most responsible party for local environmental pollution, 62 percent of the respondents chose local governments as the most important one responsible for local environmental pollution, suggesting the strong characteristic of "government-dependence" in public environmental consciousness. Sixteen percent thought the role should be taken by the public, and less for domestic enterprises, central government, and foreign companies and joint ventures. Less than 2 percent believed that social organizations should bear this responsibility, suggesting that most individuals are still unfamiliar with environmental nongovernment organization (ENGOs) and not understanding the role of ENGOs in environmental governance.

In local environmental governance, local governments, enterprises, and the public are the three most important actors. The public preference for these three actors is the main dependent variable to be explained in this chapter. There are two reasons to exclude central government and social organization from our analysis. First, given the research question and China's current environmental protection system, the central government is an institutional provider for local environmental protection rather than an active participant in implementation. Second, for the sake of effectiveness of statistics, the proportion of respondents selecting social organizations is too low to be included in the regression model. The analysis merged domestic and foreign enterprises and deleted the data that have no response to the dependent variable. The final dataset has 2161 explicitly expressed respondents.

2.2.2. Independent Variable: Age (Year of Birth)

According to the socialization theory, primary factors affecting social transformation include: (1) change in adult experience, such as unemployment and divorce; (2) major social events, such as economic recessions; (3) long-term social transformation, such as social transition; (4) leadership changes; (5) the change of social agents of socialization (Sigel 1989). We believe that intergenerational differences in consciousness of environmental participation reflect the social transformation. So we consider social transformation, social structure, and major historical events as factors in determining age thresholds. The population is divided into four age groups: (1) born before 1949 (the founding of the nation); (2) born between 1950 and 1965 (before the Great Cultural Revolution); (3) born between 1966 and 1979 (before the reform); (4) born after 1980 (after the reform).

2.2.3. *Control Variables*

(1) Social and Economic Variables

The social and economic variables include gender, educational level, household per capita income, and occupation in terms of type of industry including primary, secondary, and tertiary industries. These four variables all are categorical and are transformed into dummy variables in the model.

(2) Consciousness to Environmental Pollution

Studies have shown three mechanisms of public participation to environmental protection: pollution-driven, world-view, and post-materialism models (Tong 2002). The first model suggests that whether the public participates in environmental governance is based on their personal feeling of pollution severity. Thus, the public perception of nationwide environment and local environmental problems are taken as control variables. Perception of nationwide environmental problems is a continuous variable on a scale of 0–10, higher scores of which indicate severer environmental problems a person perceives. The average score is 5.77, suggesting that the public perceived nationwide environmental problems to be severe. Perception of local environmental problems is a categorical variable. The answers for it run from "very severe" (16.56 percent), "generally severe" (28.98 percent), "not too severe" (31.54 percent), to "not severe at all" (22.92 percent).

(3) Experience of Environmental Pollution

The theory of cognitive psychology suggests that there is a "recent effect" when understanding and evaluating things. To measure this effect, respondents were asked whether they experienced environmental pollutions in the past year. The 2008 survey data show that 32.35 percent of the respondents had this experience.

(4) Evaluation to Government Performance

Since the local environmental protection is region-separated and system-dependent, we choose public evaluation on the performance of local government at two levels: the city/county level and the rural villages/urban community level. The reliability test and factor analysis (not shown here) give the reliability coefficient of the two 0.62, the factor loading rate 0.7071, suggesting a significant correlation and the same dimension of them. So, the sum of these two variables can establish a comprehensive evaluation indicator of the public evaluation to local government performance. After the data are merged, the variable becomes a continuous one with a range of 0–20, a sample mean of 14.14.

2.3. Data Sources

The analysis relies on two sources of data. One is from RCCC's "civil culture and harmonious society" survey. The survey was conducted in 2008 in 73 county-level units across 25 provinces, municipalities, and autonomous regions, and completed a valid sample of 3898 Chinese citizens over 18.

In order to cover floating population, the survey used GIS/GPS-assisted area sampling method and got a probability sample with stratified and multistage probability proportional to size (PPS) methods.

To compare the changes of public environmental consciousness and analyze the performance of government-led public participating institution, this analysis also uses the data of "national public environment consciousness survey," commissioned to the RCCC by the Ministry of Environmental Protection and Education and was designed and implemented in 1998–1999. The survey was conducted in 139 county-level units across the country and the sample size was 9202. What is worth mentioning is that this survey specifically contained questionnaires for teenagers. The final qualified adult sample is 9919, and the teenager one is 2682.

2.4. Model Specification

Both domestic and foreign scholars have done research in public environmental consciousness, but quantitative research based on large-scale social surveys is still rare. Quantitative research methods not only overcome the dilemma of normative analysis and qualitative research on representation and objectivity but also effectively test the net effect of independent variables on dependent variable. Meanwhile, this study also uses the method of cohort. It compares the two nationwide surveys of a longitudinal span of 10 years, which is more scientific in cause-and-effect analysis from a methodological point of view.

According to the structure characteristics of the 2008 survey data, we select Multinomial Logistic Regression model to build a multivariate regression model of the factors related to the public preferences for environmental governance, described as follows.

$$y = \text{logit } p(y_{it} = 1, 2, 3) = b_0 + bx + b_i x_i + b_j x_j + b_k x_k + b_m x_m$$

y: dependent variable: the public preferences for environmental governance
x: independent variable: birth year
x_i: control variable group 1: social and economic status
x_j: control variable group 2: consciousness to environmental pollution
x_k: control variable group 3: experience of environmental pollution
x_m: control variable group 4: evaluation to government performance

3. Description of Intergenerational Difference on the Public Preferences for Environmental Governance

Table 10.1 shows the regression results based on the model described above. It suggests that, after controlling social and economic status, perceptions of environmental pollution, experience of environmental pollution, and evaluation of government performance, age has a significant impact on

Table 10.1 Multinomial logistic regression model of public preferences for environmental governance

Dependent variable: The public preferences for environmental governance	Public/local government (base outcome = local government)	Public/entrepreneur (base outcome = entrepreneur)	Standardized regression coefficients
Independent variable: Birth age differences			
Birth age: 1949 and before			
1950–1965	1.402+	1.364	0.087+
1966–1979	1.711**	1.885*	0.139**
1980 and after	2.505***	1.599	0.187***
Control variable group 1: Social and economic status			
Education level: Primary and below			
Middle school	0.891	1.033	
High school	0.762	0.889	
College and higher	0.675	1.414	
Female	1.511***	1.449*	0.113***
Villages	1.576**	1.198	0.117**
Per capita household income: 2,000 and below			
2,000–4,000	1.024	1.515	
4,001–6,999	1.002	1.306	
7,001–9,999	1.193	1.351	
10,000 and higher	0.724	0.799	

Industry type:		
Secondary industry		
Primary industry	1.036	2.005**
Tertiary industry	1.305	1.486
Control variable group 2: Consciousness to environmental pollution variables		
Perception of nationwide environment	1.004	1.024
Perception of local environment: Very severe		
Generally severe	1.729**	1.303
Not too severe	1.588*	0.817
Control variable group 3: Experience of environmental pollution variables		
Experienced environmental pollutions in the past year	0.750*	0.640*
Control variable group 4: Evaluation to government performance variables		
Evaluation to local government performance	1.047***	0.994
	0.147**
	0.124*
	0.124***
LR $\chi^2(40) = 180.61$***		
Pseudo $R^2 = 0.0487$		

Notes: The data in the table are odds ratios; $^+$means $p < 0.1$, * means $p < 0.05$, ** means $p < 0.01$; *** means $p < 0.001$.
Data Source: Civil Culture and Harmonious Society Survey, 2008, by RCCC.

Table 10.2 Intergenerational difference in public preferences for environmental governance

	1949 and before	1950–1965	1966–1979	1980 and after	Total
Enterprises	16.05	14.37	12.85	18.52	14.85
Local governments	71.85	69.67	67.32	57.26	67.28
The public	12.1	15.97	19.83	24.22	17.86
	(405)	(689)	(716)	(351)	2,161

Data Source: Civil Culture and Harmonious Society Survey, 2008, by RCCC.

public preferences for environmental governance. The standardized regression coefficients can be used to describe the size of each independent variable to the dependent variable. The data show that the coefficients for age groups are statistically significant and their size is larger than that of any other control variables, suggesting that age is the most important factor among others affecting the public preferences for environmental governance.

A longitudinal observation of preferences in intergenerational differences, as is shown in Table 10.1, suggests that citizens born in different periods show an incremental difference on selecting "Government" or "public," which is significant. From Table 10.1 we can see that, as the age increases, the option for "local government" is far higher than "public" in odds ratio. The "public" option odds ratio for group born after 1980 is 2.505 times more than the option for "local government," and the standardized regression coefficient is also the highest of 0.187, passing the test of significance at $p < 0.001$. The descriptive results also show the intergenerational differences: the group born after 1980 has the highest preference for "public" among other groups, which is 24.22 percent. The pattern in Table 10.2 suggests that the younger the generation, the more the preference for "public" governance; the elder the generation, the more the preference for "local government."

4. Explaining the Intergenerational Difference in Public Preferences for Environmental Governance

4.1. Policy Changes and Social Mobilization of Environmental Protection

The Initial Stage (1972–1978). The Chinese government began the social mobilization of environment in 1972. At that time, the government put forward the "thirty-two-word(s) policy," which emphasized the importance of the public in environmental protection (Hong 2001).

The Pollution Control Stage (1973–1979). The government put the prevention and control of environmental pollution as the main task of environmental education. But society-wide education and mobilization did not take effect.

The Environmental Management Stage (1979–1992). Environmental education took effect, during which environmental management became the main goal of environmental education.

The Environmental Education Stage (since 1993). Environmental education was adjusted to meet the demand for sustainable development. Higher-level environmental education developed substantially and rapidly, while the primary-level continued to improve. Special courses of environmental education were offered in primary and middle schools. The propaganda departments carried out environmental publicity activities under the theme of "centurial trip of Chinese environmental protection." The ENGOs, such as the Friends of Nature, formed a new force of environmental education. The environmental humanities featured by environmental ethics became a part of the curriculum in universities.

4.2. Socialization in Environmental Consciousness among the Age Groups

According to the early classical socialization theory, the socialization process of an individual occurs mainly at an early age before reaching adulthood, normally between 10 and 15 years. From the life course perspective in the political socialization theory, this period is between formation of abstract allegiance to the nation and government and initialization of citizenship (Riccards 1973).

Reflecting on the evolving environmental protection system in the past decades in China, one can see that the generation born before 1949 did not receive much environmental education. Neither did the individuals born between 1950 and 1965, even though the environmental protection was put forward by the government. But the lack of actual education and public participation during their teenage in the 1960s and 1970s made them less aware of the importance of environmental protection. Significant differences in environmental consciousness occurred among the people born between 1966 and 1979. The environmental education was in initial stage from 1978 to 1993, a period when this cohort consisted of teenagers. Since the environmental education at this stage was mainly implemented in higher education, people with a higher educational background received more environmental education, and the education they received were more professional and specific. Individuals at the lower educational level, in contrast, exposed themselves to much less environmental education. The situation became much different for the generation born after 1980. When they spent their teenage years in the 1990s, the public environmental education witnessed a rapid development. This generation received their environmental education in their primary and middle schools. In addition, the EGNOs began to promote public environmental education along with the government during this period.

In support of this argument, Figure 10.1 shows a strong correlation between the level of environmental consciousness among different age groups and policy stages during which they spent their teenage years. For example, those born before 1949 and spent their teenage years during the stage with no environmental education are the least possible group to choose "the public" (12.10 percent) as local environmental governance. However,

Level of environmental consciousness	12.10%	15.97%	19.83%	24.22%
Age cohort	Born before 1949	Born between 1950 and 1965	Born between 1965 and 1979	Born after 1980
Policy stage of environmental education	Before the initial stage	The beginning stage	The initial development stage	The rapid development stage

Figure 10.1 Environmental consciousness, age cohort, and environmental education

Note: The figures in the table refer to the percentage of respondents of born in different years that chose "the public," when they were asked about the most important in regional environmental governance in the 2008 survey.

those born after 1980 and spent their teenage years during the rapidly growing stage of environmental education have the highest percentage of choosing "the public" (24.22 percent).

This result suggests that the environmental education in one's teenage years has a significant impact on their later environmental consciousness, and the task is mainly undertaken by primary school. So, we can see that environmental education has actual effects rather than theoretical speculation.

4.3. A Micro-level Explanation Based on Temporal Comparison

4.3.1. The Preferences for Environmental Governance of the People Born after the 1980s
As previously discussed, cultivating the public environmental consciousness is the goal of the public environmental participation system. We have argued that, compared to those who choose "enterprises" or "local government" as environmental governance, those who choose "the public" have a higher environmental consciousness and will become the driving force in China's environmental participation system. The analysis above has confirmed this argument. As is shown in Table 10.2, nearly a quarter of the individuals born after 1980 chose "the public" option for environmental governance, approximately 7 percent higher than the average across all age groups. This implies that this group of individuals has the strongest tendency for environmental participation.

In order to study the characteristics of specific group's environmental consciousness, we need to introduce the political socialization theory. Usually, individual behaviors are attributed to both historical and contemporary factors. The political socialization research is to explore the historical factors on individual's contemporary political attitude and behavior (Sears 1975). This offers us a useful perspective. We can date back to one's adolescent years, the critical period of one's socialization, and make a comparison with the data of their current environmental consciousness. The data of 1988 and 2008 surveys provide us with such an opportunity to explain the socialization process of environmental consciousness and its effect on institutional change.

4.3.2. Teenagers in the 1988 "National Public Environmental Consciousness" Survey
The data of the 1988 National Public Environmental Consciousness survey show that teenagers with age between 10 and 15 years have much stronger environmental consciousness than adults. First, the teenagers are more concerned with environmental protection in contrast with adults. In ranking the problems facing China, teenagers put environmental protection at the first position among six choices (Environmental problem, Education deficiency, Poverty, Population, Natural disaster, and Regional war), while adults put it in the fourth place next to poverty, education, and poverty. This suggests that teenagers are more concerned with environment.

Second, from environmental behavior perspective, the percentage of teenagers choosing environment-friendly behavior is nearly 10 percent higher than adults. More teenagers (37.8 percent) chose the environment-friendly behavior in dealing with trash, such as "bring home," "take them on," or "take them until find a dustbin," than adults (29.4 percent).

Lastly, the school education is the first and most important source of their environmental knowledge. Teenagers overwhelmingly pointed to school (70.30 percent) as the most important source of their environmental knowledge, compared to television (54.1 percent), newspaper (28.9 percent), parents or other family members (18.7 percent), and friends (6.7 percent). This suggests that primary and secondary education plays a critical role in cultivating teenagers' environmental consciousness.

4.3.3. Teenagers' Socialization Process of Environmental Consciousness and Their Consciousness in Environmental Participation

The teenagers whose age is between 10 and 15 years in the 1998 survey are exactly the groups born after the 1980s in the 2008 survey. We can find that higher environmental consciousness in teenage years do help them develop their consciousness of environmental participation after growing up. Because they have received the most systematic environmental education during their teenage years, their level of environmental consciousness can be a good indicator of the performance of the government-led environmental participation system because school education has played a key role in the cultivation of teenagers' environmental consciousness and its translation into group consciousness shared by the entire generation. The comparison of the two surveys provided us with strong evidence to make two statements: First, as the media of socialization, the primary and secondary school education plays a key role in the public socialization process of environmental consciousness. Second, socialization of environmental consciousness improves the tendency of environmental participation.

We can get some further enlightenment from the analysis above. Although the current participation model is dominated by the government with a top-down approach, the socialization of environmental consciousness has nevertheless been internalized in the preferences of the post-1980 generation and made a positive impact on their tendency for participation. As a result, a bottom-up environmental participation mechanism at the grassroots level may grow. It is therefore safe to say that the environmental consciousness has gradually accumulated from individual-level consciousness into a society consensus through the socialization of the post-1980 generation, therefore facilitating the process of transition from a top-down model to a bottom-up one, or, more likely, a complementary model with both government and grassroots efforts working together to better encourage public environmental participation and improve environmental governance.

5. Conclusion

Through the comparison of the data across a decade, one of the key findings here is that the teenagers of the 1980s have a higher environmental consciousness than the adults at the same period. And 10 years later, when they became the group of the "post-80s" in the 2008 survey, they showed higher consciousness of environmental participation than those of other age groups. The socialization theory explains this phenomenon well: individual-level environmental consciousness has been translated into a social consensus through the socialization process of the post-80s, which dominates their preference regarding the environmental governance.

In studying the socialization of the teenager's environmental consciousness, we find that social changes are the most important factors in individuals' socialization. Environmental issues in China emerged as the result of modernization and urbanization, with which a significant outcome is the interruption of the normal socialization process. Intergenerational differences thus follow. Special social events have a similar mechanism. Under the government-led environmental participation system, the popularization of environmental education became such a special event for all the post-80s. That special event played an important role in transforming environmental consciousness from individual level to society level.

From the socialization perspective, the environmental education has two modes of socialization. The first is the direct effect of the top-down environmental education. The environmental education implemented by the Chinese government effectively enhanced the environmental awareness of the teenagers. Second, the environmental education received in one's teenage will gradually be internalized to make them aware of environmental governance, driving the perception of environmental governance toward a more diversified direction from the single government mode in a bottom-up one. With a combination of these two modes, environmental education has produced both short-term and long-term institutional effects.

Our analysis of intergenerational environmental consciousness suggests that the process of socialization has helped the government to achieve its goal to a certain extent, including cultivating the public consciousness of environmental participation and promoting the diversification of environmental governance. This process not only helps improve the performance of environmental governance but also facilitates the transition from a top-down model of governance to a bottom-up one. Based on such an assessment, we can expect the surge of a new force of public environmental participation that, by diversifying the model of environmental governance, may successfully constrain the impetus for a GDP-dominated growth model adopted by most local governments and make sustainable development possible in the future.

Bibliography

Hong, Dayong. 2001. *Social Transformation and Environmental Problems* [*shehuibianqian yu huanjing wenti*]. Beijing: Capital Normal University.
Moschis, George P. and Gilbert A. Churchill. 1978. "Consumer Socialization: A Theoretical and Empirical Analysis." *Journal of Marketing Research* 15: 506–601.
National Environmental Protection Bureau. 1988. *Selected Environmental Protection Papers* [*huanjingbiaohu wenjian xuanbian*]. Beijing: China Environment Press.

Pan, Yue. 2004. "Environmental Protection and Public Participation—A Lecture Given on World Environment Celebrity Reports on Scientific Development [*zai kexuefanzhanguan shjie huanjing mingren baogaohui shang de yanjiang*]." Available on the website of the General Office of Environmental Protection Ministry, June 1. http://www.mep.gov.cn/gkml/hbb/qt/200910/t20091030_180620.htm

Riccards, Michael P. 1973. *The Making of the American Citizenry: An Introduction to Political Socialization*. New York: Chandler Publishing Company.

Sears, David O. 1975. "Political Socialization". In Greenstein, Fred I. and Nelson Polsby (eds.) *Handbook of Political Science, Vol. 2*. Reading, MA: Addison-Wesley.

Sigel, Roberta S. 1989. *Political Learning in Adulthood: A Sourcebook of Theory and Research*. Chicago: University of Chicago Press.1989.

Tong, Yanqi. 2002. *Environmental Consciousness and the Tendency of Environmental Protection Policy* [*huanjingyishi yu huanjingbaohu zhengce de quxiang*]. Beijing: Huaxia Press.

Weissberg, Robert. 1974. *Political Learning, Political Choice, and Democratic Citizenship*. New York: Englewood Cliffs.

CHAPTER 11

CHINA'S ENVIRONMENTAL CRISIS
AND CONFUCIANISM: PROPOSING
A CONFUCIAN GREEN THEORY TO
SAVE THE ENVIRONMENT

Joel J. Kassiola

The Master said: "Both keeping past teachings alive and understanding the present—someone able to do this is worthy of being a teacher."

Confucius, *Analects*, 2.11; Slingerland 2003:11[1]

1. INTRODUCTION: WHY NON-WESTERN POLITICAL THEORY AND CONFUCIANISM NOW

The goal of this chapter is to explain how consideration of Confucius's philosophy and the two-and-a-half millennia and evolving tradition of commentary and development of Confucian thought can advance environmental political thinking and policy. I will maintain that Confucianism can provide an intellectual framework for changing China's and the world's current unsustainable path to more effective environmental thought and decision-making. This objective may provoke fundamental skepticism that needs to be addressed at the outset: why prescribe an ancient Chinese philosopher who is not part of modern Western thought nor associated with the environment? Why deviate from the longstanding practice of Western exclusivity regarding the environment and examine Confucius's and his followers' thought regarding this important topic?

In order to address these questions, I recommend that we note the recent creation and growth of a new subfield within political theory and political science that is labeled as "Comparative Political Theory" or "Non-Western Political Theory" and consider its potential importance to the mission of achieving a sustainable and just social order.

This line of new thinking argues that the advent of globalization in the twentieth century and its intensification in the twenty-first have made it clear that the core assumptions in the Western thinkers from Plato to Rawls are misguided and narrow.[2]

The editors of these first books in the new academic field provide insight into the justification for this discussion of the contribution Confucianism can make to thinking about the environmental crisis in China and the world. One of them, while recognizing the value of traditional Western political theory, quickly adds: "... in the age of rapid globalization, confinement to this [Western] canon is no longer adequate or justifiable. In our time, when the winds of trade spread not only goods but also ideas and cultural traditions around the globe, confinement to the Western tradition amounts to a parochial self-enclosure incompatible with university studies" (Dallmayr 2010:ix). Another leader of this emerging academic subfield argues as follows, again providing a rationale for this chapter's claim of the value of considering Confucianism in the face of our environmental crisis.

> There is mounting evidence which suggests that the claims of universality made by modern western political philosophy are being questioned by other cultures, or at least by the significant representatives of these cultures. Indeed, in the West itself the claims of modern western philosophy are being questioned by those who challenge the assumptions underlying modernity. Such critical inquiry makes the comparative study of political philosophies both opportune and intellectually satisfying.
>
> (Parel 2003:11)

If the globalized world today has become Marshall McLuhan's "global village" with global telecommunications, culture, environment, and economic interdependence, then *all* political thinkers and cultural traditions should be considered as sources of political insight and direction within a comparative discussion and assessment as China and the world confront our many contemporary and interconnected problems, foremost of which is the crisis of the Earth's environment. Therefore, widening the net of intellectual resources beyond the standard Western political canon seems strongly advisable for the desirable and effective environmental political theory of the future. This is especially true when the theorists are non-Western and ancient, like Confucius, whose thought may be viewed as a powerful antidote to the root cause of the environmental crisis: modernity and its values, constituting the currently hegemonic worldview understood as a complex set of ideas and social institutions.

The founders of the new subfield of Comparative Political Theory make it clear that the driving force for their innovative approach is to criticize and offer superior alternatives to the dominant modern social order and its underlying philosophy as "a neutralizing antidote to the 'baleful' power and influence of uncontested modern western political philosophy" (Parel 2003:28).[3] The merit of including Confucius and the tradition of Confucian thought in such an enterprise seems clear. Although Confucius pronounced his views thousands of years ago during a period of distinctly non-Western Chinese culture, the ongoing tradition of Confucian philosophy may be viewed as both a powerful critique of and an alternative to modernity, especially regarding the latter's essential belief in the separation and superiority of humanity to nonhuman nature, or anthropocentrism.[4] In addition, Confucianism can provide a system of ideas that can become important to the crucial but difficult process of environmental movement in the past 40 years: creating a substitute and superior worldview to the "baleful" hegemonic modernity. Confucius's profound insight about the value of the past to our understanding of the present (see epigraph), while in conflict with modernity's "contempocentric" (Speth 2008:xvii) preoccupations, becomes, I submit, an instructive alternative teaching for today in China and beyond.

Another virtue of Comparative Political Theory and analyses across time and space comparing Western and non-Western perspectives is how similarities and differences between philosophies expose presuppositions that are uncritically accepted or taken-for-granted in the modern Western view of humanity, politics, and society, particularly concerning the important relationship between nature and humanity. This comparative process, illustrated below, can improve understanding of one's own philosophy as well as the philosophy of the other tradition as a result of the comparative analytical project.

Yet, even if we accept the reasoning of Comparative Political Theory, why select Confucius and the millennia-long Confucian philosophical tradition built upon his thought? As I have argued elsewhere (Kassiola 2010:195–218), there are special qualities and beliefs in Confucian thought that make it particularly relevant and valuable—perhaps uniquely so—at the present time of dangerous environmental threats to all inhabitants of our planet. It is the aim of this chapter to defend this wide-ranging claim and to demonstrate the value of non-Western comparative environmental political theory, particularly Confucianism, and to stimulate others to conduct further similar inquiry.

2. Some Preliminary Points about Confucianism

Let us envision an alternative social order and social values to the hegemonic and unsustainable modern society that will, instead, respect the environmental limits of our finite planet. Such a postmodern sustainable society must reflect, in my view, an appreciation of the importance of morality; the central characteristic of the proposed new social order should be one whereby human beings alone are not viewed as the ultimate value as they are in anthropocentric modernity. In contrast, the Confucian social order, in its later neo-Confucian form, addresses not only the human condition but the realm of all other components of nonhuman nature, and importantly, the universe as a whole. Therefore, I advocate a Confucian Green Theory for the twenty first century based upon the Neo-Confucian "continuity of being" (Tu 1985:35–50) theory; an "anthropocosmic" or alternatively termed, "cosmoanthropic" triad or trinity among Heaven, humanity, and the Earth (Tu 1989; Tu 1993, 1998 and Ro 1998:171).

The second term, "cosmoanthropic," seems to me to be an improvement on Tu's suggested "anthropocosmic" word as both terms attempt to show how the Confucian theory overcomes the anthropocentrism of Western modernity. This is so, I believe, because the term suggested by Ro puts the reference to the cosmos first and humanity as part of the larger universe, while the term suggested by Tu Wei-ming, "anthropocosmic," places humanity first and at the center of the universe with the highest value. In opposition to Tu, Young-chan Ro calls for a "cosmological anthropology rather than an anthropological cosmology; human beings must be understood in light of the universe rather than the human rationality or logos being imposed on the universe" (Ro 1998:186).

Confucianism is in stark contrast to modernity with regard to our current veneration of the "new and improved" and its corresponding downgrading of the "old," as expressed by Confucius's counterpoint in *Analects* 2.11 (see epigraph). The vital link between the old and the new, past and present, with lessons to be learned from the past, is a foundational belief in Confucian thought. Confucius himself looked back thousands of years prior to his own time of the sixth and fifth centuries BCE to the legendary Sage Kings of Chinese history and their ideal rule in the third millennium BCE (Yao, Shun, and Yu) as well as the "Golden Age" of the Early or Western Chou Dynasty (1111–771 BCE) led by Confucius's and China's cultural heroes: King Wen, King Wu, and the Duke of Chou (Ebrey 1999; Slingerland 2003:248–255).

These figures were the creators and rulers of the longest-lasting and most successful Dynasty in Chinese history in the Early (or Western—in geographic terms) praiseworthy stage of the Chou Dynasty, the one that Confucius referred to because it embodied Confucius's social, political, and ethical ideals. We should structure the present social order, according to Confucius, based upon understanding of the past—both good and bad. The latter could include the corruption of rulers and the resulting social disorder leading to the demise of the Shang Dynasty (1751–1112 BCE) and the moral decay and full of internal discord of his own Eastern Chou Dynasty (771–221 BCE) broken down into two distinct periods, known as "The Spring and Autumn" period (722–481 BCE), which would eventually precipitate the violent and destructive "Warring States" period (480–222 BCE) that brought down Confucius's ideal Chou Dynasty in 222 BCE some 250 years after his death following hundreds of years of violence among the States despite Confucius's own valiant efforts to save it (Chan 1963:xv).

A different translation of the *Analects* 2.11 passage (see epigraph) shows the dependent relationship of the new upon the old, and, therefore, the ineradicable value of the old, according to Confucius: "The Master said:

'A man is worthy of a teacher who gets to know what is new by keeping fresh in his mind what he is already familiar with.'" (Lau 1979:64). Therefore, according to this translation, we come to know the new by retaining and utilizing humanity's past thoughts and actions, according to Confucius, who used the Western Chou Dynasty and its Sage leaders as prescriptive models for his own philosophy and recommendations to others on how to live.

Another important preliminary point about Confucian thought concerns his "bell clapper" role. This salient aspect of Confucianism for contemporary society refers to his sounding the alarm for societal transformation amidst a violent and morally decadent society in crisis. The society in upheaval in the sixth-century BCE China that Confucius was born into is expressed in *Analects* 3.24:

> After emerging from the audience [with the Master] the border official remarked, "You disciples, why should you be concerned about your Master's loss of office? The world has been without the Way [*Tao*] for a long time now, and Heaven [*T'ien*] intends to use your Master like a wooden clapper for a bell [to awaken people]."
> (Slingerland 2003:27; Chan 1963:25)

Translator Slingerland adds about this passage the following important note: "...the border official's point is thus that Heaven has deliberately caused Confucius to lose his official position [in the State of Lu] so that he might wander throughout the realm spreading the teachings of the Way and waking up the fallen world" (Slingerland 2003:27–28). *It is this crucial "bell clapper" role of sounding the alarm and "waking up the fallen world" of Confucius's time that inspire this chapter, which seeks to draw an analogy between our own "fallen world" and the alarming environmental crisis requiring transformational change and Confucius's tumultuous time and his transformational role.* (The important Confucian concept of *T'ien* or Heaven shall be discussed subsequently.).

The Eastern or Later Chou Dynasty, which Confucius was born into and formed the social context for his thought, was on its way to full-blown civil war by the time of Confucius's death with the onset of the Warring States period in Chinese history (480–222 BCE), which terminated the 800-year reign of the Chou Dynasty.

We can find similarities between Confucius's time of upheaval during the sixth and fifth centuries BCE in China, which was characterized by social corruption, disorder, a crisis of values, and violence caused by venal and decadent kings and their courts, and our modern Western consumer society (now global, including China) plagued by an environmental crisis brought on by erroneous and corrupt social values and institutions that resist and obstruct the necessary societal transformation.

Today, the world is beset by value crises and multiple wars within and between different States with the possibility of even more warfare over material resources (like oil or water) as they become scarcer, plus rampant corruption among political leaders throughout the world (including the United

States). These phenomena occur in both developed and developing nations with social unrest, value disharmony, and conflict. Decadent and environmentally degrading and unsustainable luxurious consumption echoes Confucius's chaotic time in China during the Spring and Autumn, and Warring States periods of the Later Chou Dynasty. Furthermore, our current disorganized and violent global society faces lethal threats to its existence in the form of environmental challenges from climate change, desertification, water scarcity and pollution, air pollution, soil erosion, and so on (Speth 2008: 17–45).

How our own declining global order will meet its own violent demise, whether from military conflict, like Confucius's own Chou Dynasty, or from an anthropogenic environmental catastrophe, remains to be seen. I propose that we follow Confucius's example of sounding the warning alarm to his own declining society for the need for transformation and moral improvement so many centuries ago. By examining the philosophical tradition that built Confucius's ideas, we can learn about ways to avert an environmental disaster and lead a more moral and satisfying life by creating a sustainable and just postmodern social order. As Comparative Political Theory maintains about different cultural traditions, and as I maintain about Confucius's different temporal period, comparison of disparate political philosophical traditions in space and time not only achieves the benefit of insights from the "other" tradition (in this case Chinese Confucianism) but also provides us with a deeper understanding of our own present tradition as well.

From this point of view, by engaging in comparative political analysis of both Confucianism and Modernism, we are fulfilling the ultimate Confucian goal of self-cultivation leading to self-transformation and self-realization—the highest Confucian value—or actualizing our potential humanity given to us by Heaven (Tu 1985). For Confucians, humans need to ethically cultivate themselves by seeking knowledge—learning—endlessly, ceasing only with death, as a means to realizing our Heavenly-endowed humanity (*jen*). The upshot is to try to understand the world and develop continuously in order to improve both ourselves and the world. The similarity of these Confucian ideas to the origin of Western political theory of Socrates and Plato with their philosophical watchword of: "The unexamined life is not worth living" is striking and substantiates the comparativists' approach to studying different cultural traditions and its payoff.

As I see it, for contemporary humanity as a whole, and China specifically, analogous to Confucius's own tumultuous time, the environmental crisis is like Confucius's "bell-tolling," signaling the need for societal transformation. The alarming "message" we are receiving from the environment constitutes a warning to modern societies and those societies aspiring to become modern and endlessly consumption-based, that "business as usual" or the uncritical modern values and institutions of contemporary consumer society cannot and should not (were it, *per impossible*, be able) to continue and must be transformed quickly. To that end, let us compare critically the fundamental assumptions of Modernism and Confucianism, and then consider later neo-Confucian thought (beginning in the eleventh century CE during the Sung Dynasty [960–1279 CE]), especially its cosmology with its unique vision of relationships and interactions among Heaven, humanity, and the Earth (nonhuman nature).

3. MODERNISM AND CONFUCIANISM: A CRITICAL COMPARISON

What I propose to do in this section of the chapter is to sketch (because of space limitations) a broad outline of the major differences between Modernism and Confucianism. I realize that whole books have been written on the nature of each worldview's content and values, and furthermore, that other interpretations are possible. Uppermost in my mind is the urgent need for an alternative social order that has, thus far, eluded environmentally focused critics of the modernist status quo that I think justifies the approach taken here.

James Gustave Speth, a veteran of decades in the American environmental movement has written: "...today's environmental policy and politics offer too weak a medicine, the proper perspective on environmental business as usual must be critical and must offer proposals for deeper change...The issues require a fresh conceptualization and a new way of thinking, even a new vocabulary" (Speth 2008:xiii–xiv). This "fresh conceptualization" view holds that a more positive argument than "the doom and gloom" approach of the past 40 years of the environmental movement that has not succeeded in convincing citizens and policymakers in contemporary consumer societies to give up their (impossible) idea of a ceaselessly growing materially affluent way of life (whether realized or not) that is so damaging to other realms of life: the environment, morality, politics, the family, and so on.

The practical focus of Confucius and his followers sought realizable solutions for how to live within the actual social reality of a particular time. His principle of timeliness (*shih*) emphasized how changing social conditions drive the appropriateness of specific human actions (see Confucius's *Analects* 8.13; Slingerland 2003:82). This teaching of Confucius can be especially useful to us despite the passage of so much time since Confucius's day. The reasons for this powerful parallel with a non-Western cultural tradition in ancient times, I believe, lies in the remarkable equivalences regarding the troubled nature of our respective societies.

In modern society, we suffer from moral decay and political violence. In addition to these fatal threats that eventually brought down Confucius's Chou Dynasty, we have added, ominously, a dangerous global environmental crisis. In *Analects* 14.1 Confucius considers whether refraining from "...competitiveness, boastfulness, envy, and greed" can be considered goodness (Confucius' *Analects* 14.1; Slingerland 2003:153). Despite the many centuries since Confucius's life in China, who would deny the applicability of these very same vices today, especially as they express themselves—with social support and even instigation—in our competitive materialistic consumer society?

The current time does indeed appear "ripe" for societal transformation, and China can be a world model and leader by using its Confucian heritage in constructing the needed alternative social order. In fact, it might be humanity's last great opportunity for such a transformation before some enormous disaster, like the extreme consequences of climate change, befalls the planet.

With that goal in mind, I would like to present a sample comparison between the profound philosophical building blocks of Modernism and Confucianism by the means of a chart. Of course, each of these values and doctrines can be elaborated upon at length. Nonetheless, my hope is that the summary chart of the doctrinal comparisons between Modernism and Confucianism will suffice as a starting point for the readers of this chapter and constitute an agenda for future thinking and research (Table 11.1).

I hope this analytical skeleton will give readers a general idea of the breadth and depth of the many points of opposition between Modernism and Confucianism. Each of these factors is an important component of a society's

Table 11.1 A critical comparison of modernism and Confucianism

Modernism's fundamental beliefs	Confucianism's fundamental beliefs
A. Anthropocentrism	1. Cosmoanthropism [Anthropocosmism]
Humanity's domination of nature; separation/alienation of humanity from nature with humanity superior; loss of sacred view of the Earth and nonhuman nature; humanity is the center of the universe; nonhuman nature is the means for human ends.	A cosmic trinity with Heaven, humanity, and the Earth interrelated and unified; sacredness of the Earth reinforced through reaction by Heaven; the unity of haven and humanity; and humanity and the Earth.
B. Individualism	2. Social definition of humanity
The isolated individual human being is the most important unit of analysis, and a real entity; denial of society as an independent super-individual (see Margaret Thatcher's denial of society); individuals not related to each other or nonhuman nature except self-interestedly.	Humans are essentially social-relational emphasizing key relationships: parental, spousal, sibling, friend, and ruler relationships. No conflict between self and society since the two are related and that the individual self is realized through social development
C. Materialism/Consumerism	3. Non-materialism
Human happiness is achieved through material acquisitions, endlessly and limitlessly; importance of public image and external goods are emphasized; seeking wealth is top goal.	Interior moral development is most important; materially rich life and wealth-seeking are subordinated to nonmaterial pursuit of self-cultivation, self-transformation, and self-realization.
D. Dualistic view of the world	4. Continuity of being
Oppositions between competing pairs emphasized; for example, self vs. society, matter vs. spirit, culture vs. nature, etc.	No such modern dualisms but part of unified world.

E. Economism	5. Materially simple but non-materially rich life
Economics and economic values are most important; economic values overrule nonmaterial and other internal values; supercedes other fields like politics and religion.	Nonmaterial values and life prioritized, such as: loyalty, courage, benevolence, etc.
F. Value skepticism/Value relativism/Value subjectivism	6. Value objectivism/Value realism
Skeptical of all normative discourse: politics, morality, aesthetics and theology; denies normative discourse rationality; values can only be grounded in different cultures, individuals, or paradigms.	Supports existence of the Good, which can be known by humanity; considers normative discourse rational.
G. Political reductionism/The nature of politics	7. Moralism
Separates politics from ethics; reduces politics to self-focused materialist ends.	Robust relationship between morality and politics with morality most important in society; cosmic politics seeks to achieve realization of Heavenly endowed humanity
H. Instrumental rationalism	8. Intrinsic values
Human reason emphasizes means to given ends with latter not amenable to rational discourse.	Choice of ends that are intrinsically valuable is the goal of self-cultivation and self-transformation through learning.
I. Liberalism	9. Substantive goals of the state
"Thin state," relinquishes seeking of the good; focus on procedures and procedural justice and individual rights against the state and fellow citizens.	Includes goodness; "thick state" seeking to achieve substantive values for its citizens.

Data Source: Drawn from Kassiola (2010:210–212); Kassiola (1990); and Kassiola (2003:231–234).

paradigm or view of the world, and together show the profundity and comprehensiveness of the differences between the modern and the Confucian traditions. Furthermore, by this fecund contrast with Confucianism, if the modern social order is unsustainable because of environmental limits and its normative deficiencies, then the Confucian model stands ready to inspire and shape the necessary reconstruction of modern social life through the rediscovery and reinterpretation of the Confucian Way: life based upon different political, moral, spiritual, and social truths.

In this context of the alternative possibilities of Confucianism I would like to introduce perhaps the most important aspect of Confucianism's contributions to environmentalism: the cosmoanthropism of Confucianism in which Heaven plays a central role in forming the cosmoanthropic unity between humankind and the Earth. I believe that by understanding the importance of the neo-Confucian cosmology, in sharp contrast to Western modern anthropocentrism, we can make sounder judgments regarding the health of the environment and humanity based on our connection to nonhuman nature instead of our disconnection and alienation from it.

4. NEO-CONFUCIAN COSMOLOGY AS A CONTRIBUTION TO CREATING A SUSTAINABLE AND DESIRABLE POSTMODERN SOCIAL ORDER BY ECLIPSING MODERN ANTHROPOCENTRISM

The first point that needs to be made regarding the purported contribution of neo-Confucian cosmology to a sustainable and desirable postmodern social order concerns the nature of modern anthropocentrism and its essential position within the modern paradigm of the world. Largely stemming from just a few chapters in the Judeo-Christian Scripture (Book of *Genesis*, Chapters 1–2) (Leiss 1974; White 2001), our modern anthropocentric view holds that humankind is divinely separated from nonhuman nature *and* is superior to it by divine commandment. According to this foundational view of Western Civilization, humans are empowered to control nonhuman nature for our own uses and interests.

If modernity is the main cause of the environmental crisis, as I believe it is, and anthropocentrism is at the core of modernity, then an alternative vision for this modern paradigmatic position will be necessary if we are to achieve a transformation of modernity and avoid an environmental catastrophe. Here is where we find the quintessential contribution to environmentalism by Confucianism in its neo-Confucian form.

It is important to recognize that neo-Confucianist cosmology is central to the overall worldview of the neo-Confucians, given that cosmology was not a major focus of Confucius in *Analects*, in which the Master was mostly concerned about ethics and religion. In contrast to our Earth-bound, modern view, the major significance of the neo-Confucian cosmoanthropic view of environmentalism is that one must define the ecological issues on Earth from a "cosmocentric" perspective of the universe. As one neo-Confucian scholar importantly puts it: "The earth has to be understood in relation to the universe. Ecology, in this sense, has to be situated within the proper context of cosmology: we cannot develop a proper ecology without a relevant cosmology" (Ro 1998:171).

Professor Tu, in an often-reprinted essay, "The Continuity of Being: Chinese Visions of Nature" (1985:35–50), subordinates the point that the difference between the Western and Chinese cosmologies is based upon the lack of creation myth like the one in the Book of *Genesis*. Importantly for this discussion he concludes: "The real issue is not the presence or absence of creation myths, but the underlying assumption of the cosmos: whether it is continuous or discontinuous with its creator... It is not a creation myth as such but the Judeo-Christian version of it that is absent in the Chinese mythology" (Tu 1985:35–36).

Therefore, it is the neo-Confucian belief in the continuity (or connection) with nonhuman nature along with the creator, Heaven (*T'ien*), that distinguishes it from the Judeo-Christian conception of the creation of the world by a discontinuous, transcendental God, a divine being who in the *Genesis* account creates the world *ex nihilo*, from nothing. It is within this fundamental difference regarding the nature of the universe that the sharp contrast between Western and Chinese cosmologies emerges. We are presented with a universe created by an external disconnected (divine) source (Judeo-Christian) versus one in which the universe exists in an ongoing continuous and endless process of creation, producing a unified cosmological theory (Confucian). As Tu puts it: "there is no temporal beginning to specify, no closure is ever contemplated" (Tu 1985:39).

In this Chinese cosmological view, the central idea of the cosmoanthropic trinity among Heaven, humanity, and the Earth, as mentioned in *The Doctrine of the Mean*, is most important:

> Only those who absolutely sincere [*ch'eng*] can fully develop their nature. If they can fully develop their nature, they can then fully develop the nature of others. If they can fully develop the nature of others, they can fully develop the nature of things. If they can fully develop the nature of things, they can assist in the transforming and nourishing process of Heaven and Earth, they can thus form a trinity with Heaven and Earth.
> (Chan 1963:107–108)

Translator Wing-Tsit Chan comments on this significant passage as follows: "The important point is the ultimate trinity with Heaven and Earth. It is of course another way of saying the unity of man and Heaven or Nature, a doctrine which eventually assumed the greatest importance in Neo-Confucianism" (Chan 1963:108).

Heaven, which is not a place but "the source of normativity in the universe" (Slingerland 2003:239; Eno 1990:81–82), endows both humanity and the Earth with their respective natures, and, thus, creates a triad of relationships among these cosmological entities. Implied in this cosmological theory that encompasses both humankind and the Earth, is the unity of all three cosmological components of the universe. This, in turn, creates a set of bilateral relationships between Heaven and humanity, between humanity and the Earth, and finally between Heaven and Earth. Tu puts this crucial point clearly: "To say that the cosmos is a continuum and that all of its components are internally connected is also to say that it is an organismic unity, holistically integrated at each level of complexity" (Tu 1985:39).

The remarkable passage from neo-Confucian, Chang Tsai, about Heaven being humankind's father and the Earth its mother reflects the neo-Confucian cosmological trinity and makes all humanity the "children of the universe" and connected to all "things" in it as "companions."

> Heaven is my father and Earth is my mother, and even such a small creature as I finds an intimate place in their midst. Therefore that which fills the universe I regard as my body and that which directs the universe I consider as my nature. All people are my brothers and sisters, and all things are my companions.
> (Chang Tsai 1963:497)

The sixteenth-century Confucian, Wang Yang-ming, expresses this crucial point in a famous adage: "I really form one body with Heaven, Earth and the myriad of things" taken from the following important passage for our purpose here:

> Master Wang said: "The great man regards Heaven and Earth and the myriad of things as one body. He regards the world as one family and the country as one person. As to those who make a cleavage between objects and distinguish between the self and others, they are small men. That the great man can regard Heaven, Earth and the myriad of things as one body is not because he deliberately wants to do so, but because it is natural to the humane nature of his mind that he do so... Everything from ruler, minister, husband, wife, and friends to mountains, rivers, spiritual beings, birds, animals, and plants should be truly loved in order to realize my humanity that forms one body with them, and then my clear character will be completely manifested, and I really form one body with Heaven, Earth, and the myriad of things."
> (Wang Yang-ming 1963:659, 661)

The Chinese civilization is deeply characterized by respect for one's ancestors and the duty of filial piety (*xiao*) as Confucius emphasized throughout the *Analects* (according to Slingerland 2003:238, see *Analects*:1.2, 2.5, 2.7–2.8, and 19.18). Slingerland (2003:238) describes this Confucian value as follows: "The virtue of being a dutiful and respectful son or daughter, considered by Confucius to be the key to other virtues developed in later life." Therefore, a filial relationship between Heaven and Earth and all the Earthly inhabitants is of utmost importance. Consider the environmental consequences wherein humankind is duty-bound to honor, respect, and take care of its cosmic mother, the Earth. The contrast is stark here between this neo-Confucian filial view toward the Earth and the West's hierarchical anthropocentric view of domination and exploitation of the Earth as humans see fit.

The neo-Confucian cosmology engenders respect, even reverence, for the Earth as a creation of Heaven (and, of course, Heaven, as well), plus for all the "myriad things" on Earth as our "companions" that including fellow humans, birds, animals, plants, and (puzzlingly to the Westerner) even "tiles and stones." Wang Yang-ming writes that identifying with humans, birds, animals, and plants shows one body is formed understandably among living creatures. He adds, which to the Western mind would seem cryptic, "Yet even when he [humanity] sees tiles and stones shattered and crushed he cannot help a feeling a regret. This shows that his humanity forms one body with tiles and stones" (Wang Yang-ming 1963:660). This Confucian cosmoanthropic unity of the "continuity of being" to include such things as "tiles and stones" is perhaps beyond the Western modern's ken, as it is drawn from an incommensurable cosmology that unites the range of entities from stones to Heaven, including humankind and all other living creatures and myriad things.

The factor that controls all these different members of the grand cosmic unity from Heaven to Earth to humanity to plants, animals, stones, and tiles, "the myriad things," is the rich Chinese concept of *ch'i* (see

Chan 1963:495–517, 784, where Chan translates this important Chinese ontological term as "material force"). Chan goes on to explain this concept at the start of the neo-Confucian philosophy in the eleventh century CE: "*ch'i* denotes the psychophysiological power associated with look and breath. As such it is translated as 'vital force' or 'vital power...'" (Chan 1963:495–517, 784). Tu comments on this comments as follows: "The continuous presence of *ch'i* in all modalities of being makes everything flow together as the unfolding of a single process...the motif of wholeness is directly derived from the idea of continuity as all-encompassing" (Tu 1985:38). Space limitations do not permit a discussion of the Confucian philosophy of *ch'i* and its two well-known components of *yin* and *yang*. However, I do want to, at least, summarize the implications of this cosmoanthropic theory of Confucianism.

4.1. Summary of the Implications of Confucius's Cosmoanthropic Theory

1. Nonhuman nature (or the Earth) is *not* humanity's subordinate or means to be exploited for humanity's goals or values. There is no domination of nature in (neo-)Confucian cosmology or morality. And, furthermore, because under Confucian cosmoanthropism, both humankind and the Earth are equally endowed by Heaven, both entities are equivalent components of the cosmic trinity with Heaven.
2. Since the Earth (or nonhuman) nature is Heavenly endowed, humankind must treat nature with respect and care since all Earthy entities are united "companions" under Heaven.
3. Humans are neither the center of nor the lords of the universe but equal partners with the other members of the Confucian cosmological trinity: Heaven and Earth. It seems odd for us to equate Heaven with humanity but that is one of the important teachings of the Confucian cosmology. This follows from the belief that Heaven's endowment to humanity is only in potential and it remains humans' responsibility to cultivate and transform ourselves to fully realize Heaven's endowment. Confucians emphasize this as "learning to be human." In this self-fulfillment process, Heaven and humankind are co-creators of the universe—and need each other: Heaven needs humanity to realize its potential ("The Master said, 'Human beings can broaden the Way—it is not the Way that broadens human beings.'" *Analects* 15.29; Slingerland 2003:185), and humanity needs Heaven for the initial potential to be actualized through its self-cultivation and self-transformational learning process, which is endless. Humanity is a full and equal participant in the cosmoanthropic process envisioned by Confucianism. Translator Slingerland comments on this point: "As Cai Mo [Fourth century CE commentator on the *Analects*] explains, 'The Way is silent and without action, and requires human beings in order to be put into practice;' and, 'The Way thus is transcendent,'

in the sense that it continues to exist even when it is not being actively manifested in the world, but it requires human beings to be fully realized" (2003:185–186).
4. The Earth is sacred; it is endowed by the same Heaven as humanity is endowed by. Thus, we must treat it as such, with respect and care that is deserving of Heaven.

In sum, while Confucians emphasize unity between Heaven and humanity, Confucian Greens, interpreting Confucianism for an environmentally based transformation of modernity, can stress the unity between humankind and the Earth as an effective antidote to the alienating separation from the Earth of humanity by Modernism's anthropocentric worldview. *In this cosmoanthropic trinity of Confucianism, I believe contemporary humanity can find a compelling alternative conceptual framework for constructing an environmentally sustainable and morally just new world that is not based upon the misguided anthropocentrism of modernity.*

I do not pretend that this necessarily brief presentation of the Confucian and neo-Confucian cosmological theory with its central vision of a cosmoanthropic trinity among Heaven, humanity, and the Earth is anything close to the richness of details required for a full discussion (for example, it omits how the philosophies of *ch'i* and *yin/yang* provide an ontological theory of the matter or material force that all "things" in the world share), but I hope it is sufficient to demonstrate a clear contrast to the hegemonic modern anthropocentrism of the West. In this manner, I suggest how Confucianism can provide conceptual resources for understanding today's environmental crisis and help us create another path. With this accomplishment, we can truly learn from Confucius and Confucianism, and, perhaps, thereby, save the planet, ourselves, and all the "myriad things" on Earth that are our "companions."

Notes

1. Conventional notation for the *Analects* is by book and chapter such that "2.11" means: Book 2, Chapter 11. Hereafter all references to the *Analects* will follow this standard notational convention with the specific translation noted as well.
2. For recent works within this new scholarly emphasis within political theory that seek to transcend and expand previous European and American domination of the field, see Parel and Keith 2003; Dallmayr 2010; and, Freeden and Vincent 2013.
3. The word in single quotes comes from the *Oxford Dictionary's* description of the Upas tree (used in the subtitle of this book): "a fabulous Javanese tree so poisonous as to destroy life for many miles around." Metaphorically, it stands for an entity that has a "baleful power or influence;" see p. 27 of Parel 2003 for this description of the Upas tree.
4. On anthropocentrism, the central doctrine of modernity, see any text on environmental ethics, for example, Baxter 2012: 355–358.

REFERENCES

Baxter, William F. 2012. "People or Penguins: The Case for Optimal Pollution," In Barbara Mackinnon (ed.) *Ethics: Theory and Contemporary Issues, Seventh Edition with Readings*. Boston: Wadsworth Publishers, pp. 355–358.
Chan, Wing-Tsit. (trans. and compiled). 1963. *A Source Book in Chinese Philosophy*. Princeton: Princeton University Press.
Chang, Tsai. 1963. "The Western Inscription." In Chan, Wing-Tsit, (trans. and compiled). *A Source Book in Chinese Philosophy*. Princeton: Princeton University Press, pp. 497–517.
Dallmayr, Fred. (ed.) 2010. *Comparative Political Theory: An Introduction*. New York: Palgrave Macmillan.
Ebrey, Particia Buckley. 1999. *The Cambridge Illustrated History of China*. Cambridge: Cambridge University Press.
Eno, Robert. 1990. *The Confucian Creation of Heaven: Philosophy and the Defense of Ritual Mastery*. Albany: State University of New York Press.
Freeden, Michael and Andrew Vincent (eds.) 2013. *Comparative Political Thought: Theorizing Practices*. London: Routledge.
Kassiola, Joel J. 1990. *The Death of Industrial Civilization: The Limits to Economic Growth and the Repoliticization of Advanced Industrial Society*. Albany: State University of New York Press.
Kassiola, Joel J. (ed.) 2003. *Explorations in Environmental Political Theory: Thinking About What We Value*. Armonk: M. E. Sharpe.
Kassiola, Joel Jay. 2010. "Confucianizing Modernity and 'Modernizing Confucianism': Environmentalism and the Need for a Confucian Positive Argument for Social Change." In Joel Jay Kassiola and Sujian Guo (eds.) *China's Environmental Crisis: Domestic and Global Political Impacts and Responses*. New York: Palgrave Macmillan, pp. 195–218.
Lau, D. C. 1979. trans. Confucius, *Analects*. New York: Penguin Books.
Leiss, William. 1974. *The Domination of Nature*. Boston: Beacon Press.
Parel, Anthony J. 2003. "The Comparative Study of Political Philosophy." In Anthony J. Parel and Ronald C. Keith (eds.) *Comparative Political Philosophy: Studies Under the Upas Tree*. Lanham, MD: Lexington Books, pp. 11–28.
Parel, Anthony J. and Ronald C. Keith (eds.) 2003. *Comparative Political Philosophy: Studies Under the Upas Tree*. Lanham: Lexington Books.
Ro, Young-chan. 1998. "Ecological Implications of Yi Yulgok's Cosmology." In Tucker, Mary Evelyn and John Berthong (eds.) *Confucianism and Ecology: The Interrelation of Heaven, Earth and Humanity*. Cambridge: Harvard Center for the Study of World Religions, pp.169–186.
Slingerland, Edward, (trans.) 2003. Confucius, *Analects with Selections from Traditional Commentaries*. Indianapolis: Hackett Publishing Company.
Speth, James Gustave. 2008. *The Bridge at the Edge of the World: Capitalism, the Environment, and Crossing from Crisis to Sustainability*. New Haven: Yale University Press.
Tu, Wei-ming. 1985. "The Continuity of Being: Chinese Visions of Nature." In Wei-ming Tu (ed.) *Confucian Thought: Selfhood as Creative Transformation*. Albany: State University of New York Press, pp. 35–50.
Tu, Wei-ming. 1985. *Confucian Thought: Selfhood as Creative Transformation*. Albany: State University of New York Press.
Tu, Wei-ming. 1989. *Centrality and Commonality: An Essay on Confucian Religiousness*. New York: State University of New York Press.

Tu, Wei-ming. 1998. *Humanity and Self-Cultivation: Essays in Confucian Thought.* Boston: Cheng and Tsui Company.

Tu, Wei-Ming. 1993. *Way, Learning and Politics: Essays on the Confucian Intellectual.* Albany: State University of New York Press.

Wang, Yang-Ming. 1963. "Inquiry on the Great Learning." In Chan, Wing-Tsit (trans. and compiled). *A Source Book in Chinese Philosophy.* Princeton: Princeton University Press, pp. 659–691.

White, Lynn. 2001. "The Historical Roots of Our Ecological Crisis." In Pojman, Louis P. (ed.) *Environmental Ethics: Readings in Theory and Application.* Belmont: Wadsworth Publishers, pp. 13–18.

CONTRIBUTORS

Jianguo Chen is an associate professor of Public Management at North China Electric Power University. He received his PhD from Renmin University of China in 2009. During the 2007–2008 academic years, he was a joint PhD student at the Workshop in Political Theory and Policy Analysis, Indiana University. His main research interests are in energy and environmental policy and self-governance in urban neighborhoods. His work has appeared in over ten Chinese journals.

Sarah Eaton is an assistant professor in the Department of Political Science at the University of Waterloo in Waterloo, Canada. Her research interests lie primarily in contemporary Chinese political economy. In addition to projects on local environmental politics in China, she has also conducted research on China's national champions' strategy. Her work has appeared in *The China Quarterly*, *The China Journal*, *Review of International Political Economy*, *Business & Politics*, *The Pacific Review* and the *Copenhagen Journal of Asian Studies*.

Gary S. Green is a professor in the Government at Christopher Newport University. He has written extensively on wrongdoing by and against organizations, including *Occupational Crime* (Nelson-Hall 1997) and articles appearing in journals such as *The Annals*, *Criminology*, and *Criminal Justice Policy Review*. He served as Senior Fellow for Corporate Citizenship and Organizational Legal Compliance (2006–2008), The Magellan Center, Longmont, CO.

Yin Guan is a postgraduate in School of Humanities and Social Sciences, Beijing Forestry University. Her research interests are in public administration, especially green administration and environmental policy.

Joel J. Kassiola is a professor of Political Science at San Francisco State University. His previous main research interest was in environmental political theory and recently he has shifted his interest to the specific case of China's environmental crisis. He is the author of several books including *The Death of Industrial Civilization* (1990), *Explorations in Environmental Political Theory* (2003), and *China's Environmental Crisis: Domestic and Global Impacts and Responses* (2010). He currently serves as the editor for the Environmental Politics and Theory Series for Palgrave Macmillan.

Genia Kostka is an assistant professor of Chinese Business Studies at the Frankfurt School of Finance and Management. Her main research interests are in local environmental politics and policymaking in China. She has conducted extensive fieldwork on local government organization and environmental governance in both rural and urban China, focusing on institutional change and policy innovations at the township, county and municipal levels. Her work is published in *Comparative Political Studies*, *The China Quarterly*, *The China Journal*, *Journal of Contemporary Chinese Studies*, *Energy Policy*, *Journal of Environmental Policy and Planning*, and *Applied Energy*.

Xiaojun Li is an assistant professor of political science at the University of British Columbia. His research interests include international and comparative political economy, Chinese politics, and political methodology. His work has appeared in *Asian Survey*, *Chinese Journal of International Politics*, *Foreign Policy Analysis* as well as edited volumes.

Zhen Lin is a professor and Vice-Dean of Center for Eco-civilization & School of Humanities and Social Sciences, Beijing Forestry University. His main research interests are in green administration and environmental policy, construction and management of Eco-civilization in China. His work is published in *Chinese Public Administration*, *Comparison of Economic and Social System*, *China Higher Education Research*, and *Chinese Natural Dialectics research*.

Xiaoqing Liu is a postdoctoral fellow in the School of Government and a senior research fellow in Research Center of Contemporary China (RCCC), Peking University. Her main research interests are in comparative environmental politics, policy implementation, and methodology in survey research. She has conducted and participated in several national and local survey projects, focusing on environmental awareness and perceptions, attitudes toward citizenship in China, and civil culture and society harmony in China. Her work has appeared in *Chinese Social Science Digest*, *Political Science of China Information Center at Renmin University*, *Social Sciences Academic Press*, and *People's Daily Press*.

Bingqiang Ren is an associate professor in the School of Public Administration at Beihang University, China. He was a visiting scholar at Harvard Kennedy School (2008–2009). His research interests include environmental governance, public participation and civil society in China. Currently, he is working on a project on the environmental participation in China. His work has appeared mainly in Chinese journals such as *Chinese Public Administration*, *Journal of Chinese Academic of Governance*.

Huisheng Shou is a lecturer in Government Department, Christopher Newport University. His research focuses on international political economy, globalization, and public policies. His current research projects focus on China's welfare reform, corporate regulation, and rural governance. He was the co-author of *Innovation and Development: Reforms of Township Government Elections* (Beijing: Oriental Press 2001) (in Chinese). His work has appeared in the *Journal of Chinese Political Science* and edited volumes. He was the recipient of the Best Paper Award at the 2010 annual conference of the Association of Chinese Political Science, for his study on China's welfare reform.

Phillip Stalley is an assistant professor in the Political Science Department at DePaul University in Chicago, where he teaches courses on international relations, Chinese politics and environmental politics. He is the author of *Foreign Firms, Investment, and Environmental Regulation in the People's Republic of China* (Stanford University Press 2010) and his work can be found in academic journals such as the *China Quarterly*, *Global Environmental Politics*, and *the Journal of Contemporary China*.

INDEX

accountability, 33, 45–7, 49, 53, 104, 115, 119, 120
activists, 146, 157, 167
administrative level, 6, 29, 43, 44, 87
adults, 223, 224, 225
air pollution, 101, 142, 176–8, 181, 232
air quality, 96, 101, 116, 118, 142, 176–82
Analects, 230
anomie, 201, 205
anthropocentrism, 229
Apple Inc, 145, 146, 148, 152, 154, 197
Asian Development Bank (ADB), 22, 32, 36
attachment, 200, 206
authority, 3, 5, 6, 8, 13, 16, 32, 48–9, 51, 161, 167, 169, 183
autonomous power, 7

Bank of China, 35
Banking Regulatory Commission, 35
bureaucracy, 170, 183
bureaucratic fragmentation, 83
business culture, 11, 206

cadre evaluation system, 83, 87
cadre performance evaluation assessment, 44, 88, 104
cadre turnover (rotation), 83, 88–90, 103
cancer village, 142
capacity, 2, 3, 6–8, 10, 13, 53, 89, 96, 104, 119, 136, 152, 186, 203, 207
capitalism, 196, 201
carbon emission, 11
central government, 7–9, 15, 83, 87, 95, 98, 99, 104, 116, 170, 173, 177, 179, 195, 198, 215

centralization, 3
  compare decentralization
challenges, 2, 8
chemical oxygen demand (COD), 99, 116
chí, 238–9
citizen grievance system, 35, 38, 40–1, 42
  See also xingfang system
citizen involvement, 22
civil society, 12, 147, 154, 178
Clean Development Mechanism (CDM), 150–1
collusion, 24
command-control mode, 4
comparative political theory, 228–9, 232
compatible development, 9, 112, 114–16, 119–20
compliance, 21, 22, 152, 198
compliance gap, 142
conflicts, 1, 8, 9, 169
confrontation, 16, 161, 168, 169, 170, 172, 179, 183
Confucian green theory, 227, 230
Confucianism, 227–40
  see also neo-Confucianism
Confucian philosophy, 17, 229, 239
Confucius, 227, 228, 229, 232, 238
ConocoPhillips, 145, 152
consent, 6, 175
conservation, 52, 84
  see also energy conservation
constitution, 24
control of end-of-pipe pollution, 4
cooperation, 6, 40, 89, 153, 178, 183, 185, 186
Corporate Social Responsibilities, 36, 126, 152

corporate violence, 192
  see also organizational violence
corruption, 8, 52, 88, 89, 230, 231
criminology, 191
culture, 17, 95, 98, 99, 146, 207

decentralization, 2–3, 6–7, 83, 87, 170
decision making, 3, 6, 10, 13, 50, 186, 206
  see also policymaking
deficiency, 162, 170, 173, 214, 223
deliberation, 180
deterrence, 21, 22, 202–3
deterrence trap, 198
differential social organization, 201, 204, 205
  see also Sutherland
distrust, 169, 172, 205
Du Wei-ming, 230, 236–7
dual leadership, 3
dynamics, 16, 105

ecological civilization, 4, 22, 56, 175
ecological degradation, 21
economic competitiveness, 99, 113, 120
economic development, 1, 2, 7, 9, 15, 21, 53, 113–16, 118, 125
  see also economic growth
economic growth, 1, 2, 6–9, 17, 21, 83, 97, 98, 102, 111–20, 136, 150, 170
  see also economic development
economic instruments, 4
economic interests, 7, 8, 171
emissions reduction, 22
energy conservation, 22, 31, 46, 49
energy consumption, 9, 10, 31, 46, 117, 118, 119, 120
energy intensity, 85
enforcement, 15, 16, 21–4, 31–5, 40, 41, 44–5, 47, 49, 52–3, 101, 118, 119, 120, 136, 142, 143, 144, 149, 150, 172, 191, 197–8, 201, 203–5
enforcement trap, 203
entrepreneurial government, 173
entrepreneurs, 116
environmental awareness, 10, 13, 152, 225
  see also environmental consciousness

environmental conflicts, 171
environmental consciousness, 14, 16, 214, 215, 217, 221–5
  see also environmental awareness
environmental courts, 22
environmental crisis, 228, 230
environmental degradation, 5, 21, 84, 93, 142
environmental disclosure mechanism, 22
environmental education, 14, 17, 214, 220–3, 225
environmental financing, 125, 134
environmental governance system, xi, xii, xiv, xvii, 2–4, 5
environmental governance, xi, xii, xvii, 4, 5, 8, 9, 11, 12, 14–15, 112, 113, 116, 119, 120, 141, 144, 171, 175, 176, 178, 181–3, 185–7, 198, 202, 214, 215, 216, 217, 220–5
Environmental impact assessment (EIA), 13, 25, 26, 34, 36, 37, 38, 50, 51
environmental information, 13, 22, 35, 181, 182
environmental issues, 3, 186, 203, 225
environmentalists, 13, 142, 144, 175, 176, 179–80, 183
environmental labeling program, 133
environmental management, 24, 125, 126, 129, 130, 132, 133, 135, 146–54, 220
  see also environmental governance
environmental movement, 205, 229
environmental non-governmental organizations (ENGOs), 13, 38, 197, 207, 215, 221
  see also non-governmental organizations (NGOs); green NGOs
environmental participation, 16, 173, 178, 184, 213–15, 223–5
environmental performance, 125, 134–5
environmental pollution, xv, 1, 9, 10, 23, 27, 32, 36, 120, 167, 182, 191, 195, 215, 216, 217, 219, 220
  see also pollution prevention

INDEX

environmental protection, xiv, xvi, 1, 3, 4, 5, 7–9, 12–14, 16, 17, 21–3, 30, 39, 41–9, 52–3, 101, 111–14, 117–18, 120, 125, 130, 132, 134, 137, 144, 148, 153–4, 168, 171–2, 175–7, 179, 182–3, 213–14, 216
  agencies of, 3, 5, 6, 49, 53, 179–80, 182, 197, 205, 207, 213, 221, 223
  capacity of, 2, 7
  goals of, 30–1, 42–9, 52
  history of, 220–1
  improvement of, 118–19, 195
  investment on, 135, 150
  motivation for, 114
  regulatory framework, 21
  relationship between economic development and, xiv, 2, 9, 15, 111–16, 118–20
  system, 6, 214, 215, 221
Environmental Protection Administration (EPA), 21, 37
  see also Ministry of Environmental Protection (MEP)
environmental protection bureaus (EPBs), 3, 25, 35, 40–1, 50, 84, 142, 143, 202–3
environmental protection contract (target) responsibility system, 42
Environmental Protection Law, 4, 22, 24, 33, 42, 48, 53, 118, 141, 181
  see also Law
environmental protest, 142, 161–73, 183
environmental quality, 7, 26, 37, 44, 46, 50, 93, 99, 118, 181, 183
environmental regulation, 150, 153, 191, 197–8, 201–3
environmental regulatory system, 3, 5, 173
  see also environmental protection system
environmental responsibility, 144
environmental standards, 3, 7, 8, 25, 30, 113, 125, 133–7, 142, 144, 149, 150, 173, 195, 196
environmental sustainability, 125, 130, 136, 137
environmental violence, 192
  See also organizational violence

financial constraints, 113
firm ownership, 126, 133–5
  see also private firm, state-owned enterprises/firms
firm size, 125, 133–5, 195, 196
Five-Year Plan, 22, 31, 45–6, 49, 84, 125
Foreign Direct Investment (FDI), 137, 143–50
foreign invested firm/enterprise, 141–54
formal institution/rules, 204
fragmentation, 3, 83
  see also political fragmentation; fragmented authoritarianism
fragmented authoritarianism, 202
funding, 8, 40, 51, 95, 97, 99, 103, 104, 105, 106, 184, 203

GDP, xv, 6, 7, 11, 14, 22, 31, 42, 44, 45, 53, 85, 86, 87, 89, 91, 96, 98, 102, 113, 115–17, 119, 170, 225
GDP-oriented strategy, 7
generation, 14, 17, 200, 201, 206, 220, 221, 224
gilded era, 206
globalization, 2, 10–12, 147, 149–50, 228
  see also integration
global market, 196
global village, 228
governability, 170, 171–3, 181
governance deficiency, 170, 173
government performance, 45, 215, 216, 217
grassroots activists, 167
green credit, 35, 37
green GDP, 45
greenhouse gas emissions, 148
greening growth, 83, 84, 100–2
green NGOs, 175–87
Green Peace, 145, 146, 154, 197
green security, 35
green supply chain (GSC), 151–2
green trade, 35
Green Watch, 23, 149

harmony, 83, 116
hazardous waste, 116, 146

health, xiii, xiv, xv, 1, 13, 53, 111, 142, 146, 164, 172, 176, 182, 192, 197, 235
human resources, 8, 135

identity, 172
illegal approach, 169
immorality, 204, 206
implementation, 21, 22, 52, 53, 87, 142, 147–50
implementation gap, 84, 141, 144, 147, 150
implementation shirking, 87
inaction, 167, 168, 170
incentive mechanisms, 5, 114, 118
incentive structures, 8, 111, 113–16, 119–20
indicators, 5, 42, 44, 46, 113, 114, 116, 117, 120, 127, 141
individual behaviors, 223
individualism, 198
industrial competitiveness, 111
industrial pollution, 21
industry, 11, 24, 31, 90, 95, 96, 98–104, 119, 125, 143, 146, 151, 152, 153, 154, 164, 179, 186, 216, 219
informal institutions/rules, 204
information, 2, 24, 37, 43, 51, 52, 120, 162, 179–85, 197, 205
information disclosure, 37, 38, 43
information technology, 143, 145
innovations, 43, 113, 116, 125, 134, 206
Institute for Public and Environmental Affairs, 145, 146, 147, 152, 154
see also Ma Jun
institutional capacity, 83
institutional design, 3, 5, 9, 12, 15, 120
institutionalism, 120
institutionalized channels, 9, 26, 169
institutions, xvi, 3, 12, 14, 16, 17, 46, 87, 120, 191, 196, 204, 206, 228, 231, 232
instruments, 3, 4, 5, 15, 22, 23, 24, 25, 29, 31, 33, 35, 47, 49, 50, 52, 136
integration, 207
see also globalization
Interests groups, 170
intergenerational change, 14, 213

intergenerational difference, 16, 214, 215, 217, 220, 225
intergovernmental coordination, 120
international community, 2, 10, 11
internet, 13, 94, 165, 167, 180, 207
IOS 14000, 133

Jungle, 196
see also Sinclair

Law
 Administrative Coercion, 32
 Administrative Litigation, 32
 Circular Economy Promotion, 26, 48
 Civil Procedure, 28, 50
 Control, 29
 Criminal, 27, 50
 Environmental protection, 24, 33, 42, 53, 142
 Pollution Prevention and Control, 33
 Radioactive Waste Safety, 26
 Tort Liability, 25
 Water Pollution Prevention and Control, 29–30, 35, 37, 48
leadership, 3, 42–6, 49, 52, 83, 84, 88, 89, 91, 93, 95, 97–104, 141, 162, 215
legitimacy, xii–xvii, 1, 9, 111, 169, 173
local authority, 168–9, 171–3, 197
see also local government
local governance, 16, 90, 161, 170, 171, 173
local government, xiv, 2, 3, 5, 6–9, 11, 15, 16, 29, 33, 34, 35, 39, 42–6, 85, 87, 97, 99–102, 105, 113, 115, 117, 118, 120, 136, 162–5, 167–73, 179, 181, 182, 195, 203, 215, 216, 220, 223, 225
local interests, 87, 102, 170
local protectionism, 45, 120

Ma Jun, 39
see also Institute for Public and Environmental Affairs
market-based incentives, 22, 35–7, 50
market economy, 4
mass movements, 162
mechanism, 2, 13, 15, 22–6, 33, 35, 36, 40, 46, 50, 52, 53, 116, 120, 137,

161, 162, 181, 183, 197, 205, 213, 214, 216, 224, 225
Ministry of Environmental Protection(MEP), xiii, 3, 21, 36, 37, 108, 125, 141, 176, 177, 179, 181, 202, 203, 217
  see also Environmental Protection Administration (EPA)
modernism, 227–40
modernity, 228, 229
moral attachment, 200
morality, 200, 206, 229, 235
motivation, 91, 114, 185, 196, 198, 200
multinational corporations(MNCs), 136–7, 141–53, 196–7
  see also multinational firms
multinational firms, 11, 16, 196, 197
  see also multinational corporations
mutual understanding, 183

National Development and Reform Commission (NDRC), 39, 99
National People's Congress (NPC), 24, 28, 32, 33
natural resources, 32, 111
negotiation, 7, 11, 16, 162, 163, 166
neo-Confucianism, 230
  see also Confucianism
neo-Confucianist cosmology, 236
network governance, 185
noncompliance, 21, 52
non-governmental organization (NGO), 24, 28, 38, 39, 40, 50, 153, 207
  see also environmental non-governmental organizations (ENGOs)
normative structure, 204, 207
normative validation, 206
not-in-my-backyard (NIMBY), 152

OECD, 149
officials, 6, 7, 15, 24, 29–33, 37, 40–50, 52, 84, 87–9, 94, 96, 99, 116, 120, 145, 162–6, 168–72, 179, 180, 205
Olympics, 34
one-vote veto, 45, 52
openness, 16, 144, 147, 153
opportunities, 2, 12, 38, 88, 116, 171, 184–6, 199

organizational violence, 191–207
  see also environmental violence

participation, 10, 16, 33, 38, 50–1, 53, 162, 167, 173, 176, 178, 182–6, 194, 213–16, 221, 223–5
participatory governance, 13, 16
peasants, 161–2, 166–73
performance evaluation, 41, 43–5, 47, 51, 52, 113, 120, 215
petition, 13, 161, 162, 165, 167, 168, 172
Plato, 228, 232
PM2.5, 13, 16, 118, 142, 143, 175–83, 185–7
policy, xi, xiv, xv, xvi, 13, 21, 22, 35–6, 38, 39, 45, 53, 84, 88–90, 101, 104, 130, 152, 161–2, 166, 177, 179, 183, 213, 220, 221, 227, 233
policy capacity, 6
policy coordination, 85
policy designs, 15, 117
policy implementation, xi, 3–6, 8, 9, 13, 15, 47, 83–4, 87–9
policy implications, 84, 104, 112
policy instruments, 4, 22, 49
policymakers, 24, 104, 137, 185, 233
policymaking, xii, 3, 4, 5, 13, 84, 111, 168, 169, 171–3, 178–80
policy system, 4, 5, 35
political agenda, 5
political attitudes, 214
political fragmentation, 191, 204, 206
  see also fragmentation; fragmented authoritarianism
political promotion, 113
political socialization, 214, 221, 223
political structure, 204
political system, xii, xiii, 6, 12, 87, 162, 170, 191, 199, 203, 204, 206, 214
pollution control, 22
pollution heaven, 136–7, 144, 147
pollution intensity, 130, 133
pollution liability insurance, 35
pollution prevention, 3, 4, 28, 30, 37, 48, 178, 181
pollution trading permit, 24
pollution transfer, 7
poverty, xv, 9, 199, 223
power distribution, 2, 5

priority of environmental protection, 22, 45, 53, 56, 84, 85, 104, 181
private firm, 126–7, 133–6, 196–9
professional competency, 185, 186
professionalism, 178
profit, 198–200, 206
propensity, 198, 200, 201
protests, xii–xv, 1, 2, 9–10, 16, 35, 38, 41, 51, 52, 142, 161–4, 166–73, 183
public choice theory, 24
public involvement, 21, 187, 206
public participation, 33, 38, 53
public preference, 214–15, 217–18, 220
public services, 97, 182

race to the bottom, 196
Rawls, 228
reform and opening, 130
regional environmental supervision center, 22
regional variations, 127
regulatory framework, 21
renewable energy, 4
rent-seeking, 171
rightful resistance, 161, 162
rule of law, 23
rules of liability, 23
rural areas, 7, 8, 9, 169, 173
rural residents, 119, 162, 173

school education, 224
Scientific Development, 4, 42, 45
self-enforcement, 198
self-regulation, 144, 198
service government, 173
Sinclair, Upton., 196
  see also Jungle
social capital, 204
social changes, 4, 225
social conflicts, 161
social conscience, 184
social disorganization, 201, 204–5
socialization, 14, 200, 201, 206, 213–15, 221–5
social media, 180
social mobilization, 214, 220
social normative system, 198
social norms, 170, 204, 205

social responsibility, 36, 152, 153, 184, 205
social stability risk assessment, 38, 39
social stability, xv, 10, 38, 39, 142, 162
Socrates, 232
State Council, 3, 22, 31, 34, 37, 42, 43, 45, 47, 49, 175, 181
State Environmental Protection Administration (SEPA), 21, 37
state-owned enterprises (SOEs), 36, 89, 126, 130, 133, 135–6, 196
  see also firms
strain theory, 199–200
strategic emerging industries (SEIs), 85
street-level bureaucrats, 87
structural Marxism, 196
sulfur dioxide, 143
sustainability, 125, 130, 136, 137, 138, 148, 152
sustainable development, 2, 4, 44, 84, 111, 125, 221, 225
Sutherland, Edwin., 200, 205

technological spillover, 150
technology, 9, 99, 116–18, 120, 134, 136, 143, 145, 149–51, 153, 172, 214
teenagers, 217, 221, 223–5
total suspended particulates (TSP), 143
transformation, xii, xvi, xvii, 4, 8, 84, 85, 96–100, 103, 104, 162, 215, 231–4, 236, 239, 240
transparency, 13, 23, 37, 50, 51, 53, 145, 182
trust, 16, 47, 168–70, 172–3, 178, 186

unit of analysis, 198
United Nations, 3
United States Beef Inspection Act, 196
United States Environmental Protection Agency (EPA), 23, 24
United States Sentencing Commission, 194
Urban Assessment System, 43
urbanization, 10, 225
urban residents, 16, 119

violence, 10, 41, 167, 168, 173, 191–207, 231, 233

Walmart Inc, 100, 145, 148, 152, 154
waste water, 7, 30, 96, 100, 125, 134, 135, 143, 146
water consumption, 117
water pollution, 30, 33, 34, 48, 49, 93, 143, 146, 165, 186
White Collar Crime, 205
 see also Sutherland
World Bank, 149

Xi Jinping, 141
xingfang system, 35, 38, 40–1, 42
 see also citizen grievance system

Yale Environmental Performance Index, 24

Zero-growth model, 112

Printed in the United States of America